D0983259

Experimental Magnetism

Volume 1

Experimental Magnetism

Volume 1

Edited by

G.M. Kalvius,
Physik-Department E 15,
Technische Universitat Munchen, Germany

and

Robert S. Tebble,
Department of Pure and Applied Physics,
University of Salford, England

A Wiley—Interscience Publication

JOHN WILEY & SONS
Chichester · New York · Brisbane · Toronto

Library of Congress Catalog Card Number 79-50036

ISBN 0 471 99702 1

Printed in Great Britain

Acknowledgements

The editors wish to express their gratitude to the authors of the chapters and to Mrs D. Scott, who typed the manuscript, for their patience and perseverence and to the staff of John Wiley and Son for their editorial advice and assistance. They are also grateful to their colleagues at Munich and Salford, to Professor A.J. Freeman of Northwestern University and to Mr. Max Swerdlow of the USAF Office of Scientific Research for their support and encouragement. Acknowledgements are also due to those who kindly granted permission for the use of their diagrams and illustrations.

Contributors

Bacon, G.E.,
Department of Physics, University of Sheffield, England.

Ingram, D.J.E.,
Principal, Chelsea College, University of London, England.

Lee, E.W.,
Physics Department, University of Southampton, England.

Paoletti, A.,
Consiglio Nazionale delle Richerche, Laboratorio di Elettronics dello Stato Solido, Rome, Italy.

Pearson, R.F.,
Mullard Laboratories, Redhill, England.

Contents

Magnetic Anisotropy

Magnetostriction

Magnetic Resonance

Preface

This series on experimental methods in magnetism is intended to give a comprehensive description of the various techniques and methods presently in use for the investigation of magnetic materials. The variety of methods available in magnetism is truly vast. They may be divided roughly into two groups: the more classical measurements of bulk magnetic properties such as magnetization or susceptibility and the microscopic techniques such as microwave spectroscopy (NMR,ESR) which give specific information on the electronic structure of the paramagnetic ions contained in the sample. The latter methods have been significantly expanded in the last few decades by the addition of nuclear methods such as the Mössbauer effect or perturbed angular correlations. New developments in this field are still appearing. We mention as an example muon spin rotation. Probably the most powerful single method in modern magnetism is the scattering and diffraction of neutrons. It bridges the two groups since it gives information on cooperative phenomena in the magnetic material (e.g. its spin structure) and on the local properties of the magnetic ions (e.g. the spin density distribution). The large amount of data now available, particularly on microscopic properties of magnetic ions has been matched by recent progress in theory. As an illustrative example one may mention the ab initio calculations of spin densities using models which contain most of the essential features of a real system like a magnetic metal.

Many scientists are of the opinion that the essence of research in magnetism is to gain information on materials which are commonly called "magnets". While this is certainly an important aspect, particularly in view of the technological importance of magnetic materials, it is by no means the only incentive to carrying out experimental work in magnetism. Data on the magnetic properties of a material will also give information on the electronic structure of its constituents. Applications in chemistry for investigations of the nature of chemical bonds are probably the best known examples in this respect. However, magnetic measurements have even been useful in shedding new light on the properties of biological materials such as the heme proteins which contain a paramagnetic ion (iron) in their active center. Our basic aim in putting together this series was threefold. First, it should provide a reader who wishes to enter the field of experimental work in magnetism with information in some detail, on the various methods available for the investigation of magnetic properties of matter. He should be able to judge what is involved in carrying out an experiment in the way of equipment, time, labour, data analysis and last but not least on the monetary spending. Furthermore, he should be able to decide which method is likely to be the one which gives him the best results on the question he is interested in. Second, it should give a reader who is already working in a particular field of magnetism more detailed information on other methods, their power and their limitations. In particular he should then be able to relate results from other techniques to his own work. Third, it should present the necessary information to a reader, for example a graduate student, who just wants to learn more about a method he has found mentioned in a text book or an original research paper. For this reason, care has been taken to present the information in a way which can be followed by the non specialist - the engineer, chemist or metallurgist as well as the physicist. However, it is important to realise that this does not necessarily make for easy reading. It may also be appropriate to point out that magnetism, as a subject, is commonly considered to be "difficult", particularly in some of the so-called elementary concepts. For this reason the earlier parts of some of the chapters are devoted to the development of the background to the subject in some length. In others the subject is being dealt with in separate chapters so as to give the authors more scope in writing the descriptive sections.

The developments in experimental techniques in magnetism have been rapid and the results so significant that there may be a tendency to overlook the fact that improvements in methods of measurement are not in themselves sufficient to ensure progress. Many of the earlier workers in the field were artists in their experimental skills. Much of the present day success in obtaining reproducible results is due to the high purity of composition of the specimens and to the availability of good single crystals. This point is made not to detract from the significance of the more recent developments, but rather to underline the need to include in the plans for any project on magnetism the cost, in time as well as finance, of obtaining specimens of the required quality.

Due to the acceptance of the SI system (with the basic units m, kg,s,A) as the legal system of units in most countries, an effort has been made to make general use of units compatible with the SI system (see appendix A). We have largely followed the usage of Bleaney and Bleaney (1976) which may be consulted for basic information. As a general rule the magnetic field is described by the vector $\underline{B}$, commonly called the magnetic flux density or magnetic induction. It is measured in tesla ($1T=1kg\ s^{-2}\ A^{-1}$). Where convenient, the magnetic field strength $\underline{H}$, measured in Am^{-1}, has been used as an auxiliary vector. It is important to remember that in the SI system the values of $\underline{B}$ and $\underline{H}$ are not the same in free space and therefore $\underline{B}$ and $\underline{H}$ cannot simply be interchanged. Some of the manuscripts have been written by the authors in e.m.u. and the conversion into SI has been carried out by the editors. Because of this and since the greater part of the published work in this field has been in e.m.u. it has been found useful in making reference to these publications (and elsewhere in the text) to give results in both e.m.u. and SI units.

Neutron Diffraction

G.E. BACON
University of Sheffield

Introduction

Experimental measurements and experimental data may often be ascribed to one or other of two classes. Sometimes a measurement can be made very accurately and very reproducibly, but it may be difficult to interpret the data in terms of any simple physical or chemical model. Many magnetic measurements are of this type. On the other hand measurements of thermal conductivity are easily interpreted in terms of the kinetic theory but may be difficult to carry out with great accuracy. However, some techniques are capable of both great accuracy and direct interpretation and a good example of such a technique is the case of X-ray diffraction analysis to study the 3-dimensional structure of solids. Such investigations can produce a direct picture of the electron density of the 'building block' from which the solid is built up, with the result that the different chemical atoms within it can be located and identified: at the same time the size of the building block can be determined quite easily to an accuracy of 1 part in 60,000. Moreover it can be established that the atoms are not exactly stationary but take part in restricted motion about their mean positions, and the extent of this motion may be determined and it is often found to be anisotropic i.e. different in different directions. Thus X-ray diffraction may be said to give both a static and a dynamic view of the layout and structure of the solid, on an atomic scale and in terms of its constituent chemical elements. One of the aims of neutron diffraction is to give a corresponding view in terms of the magnetism of the constituent atoms in the solid. With any diffraction technique it is always regularity and order which are most prominently displayed and the departures from perfection are then revealed subsequently: it comes as no surprise therefore if emphasis is placed on the ability of neutron diffraction to detail co-operative forms of magnetism. Indeed it is largely as a result of using beams of neutrons that the idea of co-operative magnetism has expanded far beyond the original concept of ferromagnetism.
In the following discussion of the application of neutron methods we shall see an interplay of two themes, first a search for a fundamental understanding of magnetism and, secondly, the shorter-term

task of understanding particular magnetic materials. We shall find that in many cases the choice of the materials which have been examined has been greatly influenced by the first of these two themes.

Principles

It is customary to regard the study of neutron diffraction methods, both from a practical and an academic point of view, as a sequel to the use of X-rays and there are sound reasons why this is so. First, most of the concepts relating to scattering centres and diffracted beams and the interpretation of the latter in relation to regularities and irregularities of structure have been developed from researchs with X-rays, following the original pioneering discoveries of von Laue and the Braggs in 1912. Secondly, most problems involving the use of neutrons will have required some preliminary work with X-rays and it is essential that all possible efforts to obtain information should have first been made in this way, since it is only too evident that adequate neutron sources are extremely rare and exceedingly expensive. The significance of this is that neutron methods are only used when X-ray methods fail. From this point of view we are, in the study of magnetic materials, on very safe ground, since the ability to locate and to detail magnetic moments on an atomic scale is a property which is quite unique to the neutron. Nevertheless the neutron does see features other than magnetism in its passage through a solid - in particular it sees the nuclei of atoms - so that we shall have to give a wider account of the general technique of neutron diffraction before looking in detail at its application to the study of magnetism.

Experimental Outline

The foundations of our understanding of the technique will arise from a study of the principles of the process whereby neutrons are scattered by the atoms in solids. However it will be of advantage to look first at the more practical problems of obtaining suitable beams of neutrons and, subsequent to this, of deciding on the size and form in which our magnetic materials will be required if they are to be studied. Neutrons have been known since 1936 when they were discovered as a product of the bombardment of beryllium by the α-particles given off during the decay of radium. These radium-beryllium sources weighed only a few grams and provided extremely weak beams of neutrons when regarded in the light of our present requirements. Indeed an adequate source of this type for diffraction measurements would involve compressing 10,000 kg of radium into a sphere of about 10 cm radius, a quite impossible prospect. As a result adequate neutron beams for studying solids only became available when nuclear reactors were developed at the end of the Second World War. In a typical reactor neutrons are produced during the fission of atoms of uranium and, initially, they have a spectral distribution as shown in Figure 1, with an average energy of about 2 MeV. The neutrons emerge in all directions and their intensity is indicated by the 'flux' , which is the number which emerge each second through an area of 1 cm^2. In a reactor which is suitable for neutron diffraction work the flux can range from about 10^{13} to 10^{15}. These high energy, and therefore high velocity, neutrons are slowed down by successive collisions with atoms in the surrounding moderator and they emerge from the moderator with energies in the thermal region, as shown in Figure 2. In accordance with wave-mechanical theory they also have wave properties and their wavelength λ is equal to h/mv, where h is Planck's constant and, m,v are the mass and velocity of the neutron. Under normal conditions the wavelength of the neutrons at the peak of the spectrum in Figure 2 is a little greater than 1 Å (0.1 nm) a value which is admirably suited for doing diffraction measurements in solids, simply because it is about equal to the distance of separation of neighbouring atoms. However, for certain applications it is desirable to increase the proportion of either long-wavelength or short-wavelength neutrons. This can be done by locally lowering or increasing the temperature of the moderator in the region from which the neutrons are derived for the experiment. Thus we may have 'hot' and 'cold' sources, such as 10 ℓ of graphite maintained at 2000°K or 25 ℓ of liquid deuterium at 25°K which give enhancement of the number of neutrons in the wavelength range 0.4 - 0.8 Å and of wavelength greater than 4 Å, respectively.

The practical process of studying the diffraction patterns of solids consists essentially of bombarding a sample with a beam of neutrons and then observing the distribution of scattered neutrons around the sample. The interpretation of the distribution in space is

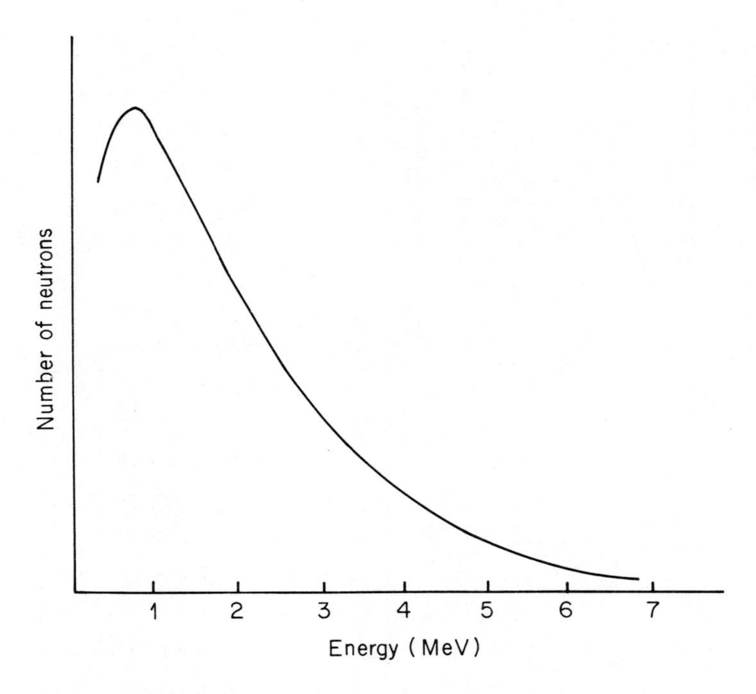

Figure 1 The velocity distribution for the neutrons produced by fission of uranium.

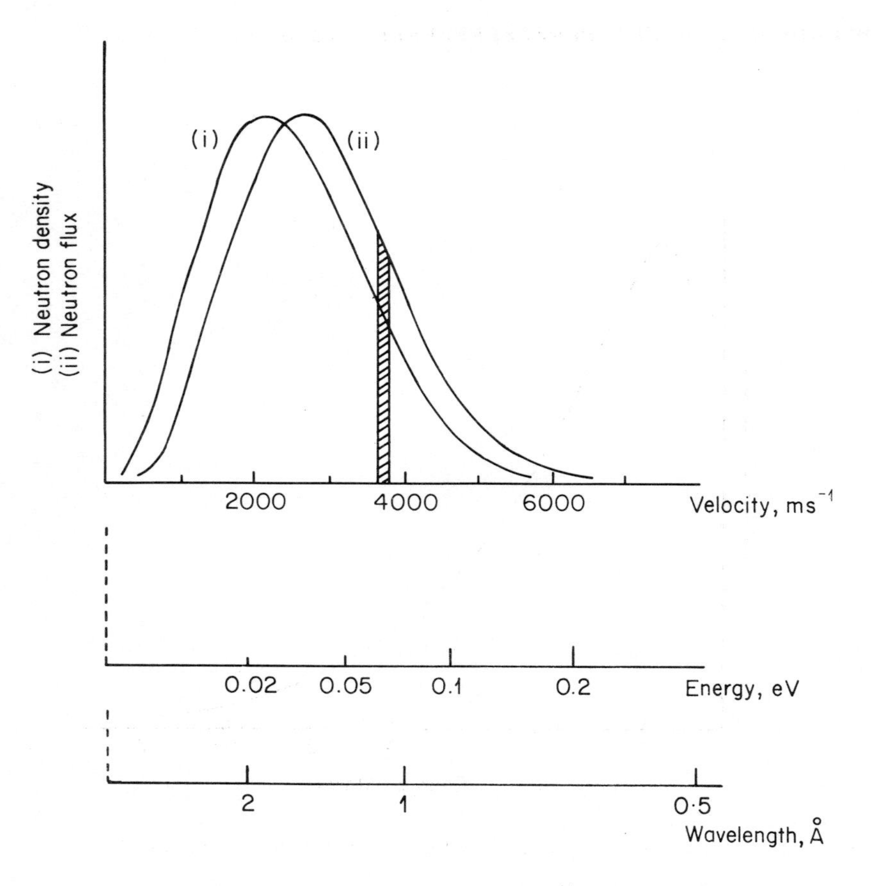

Figure 2 The distribution of velocity, energy and wavelength among the neutrons in a reactor, after moderation. Curve(i) applies to the neutrons within the moderator itself: curve (ii) indicates the distribution as the neutrons emerge from a collimator. The shaded area under curve (ii) represents a suitable choice of "monochromatic" neutrons for a diffraction experiment.

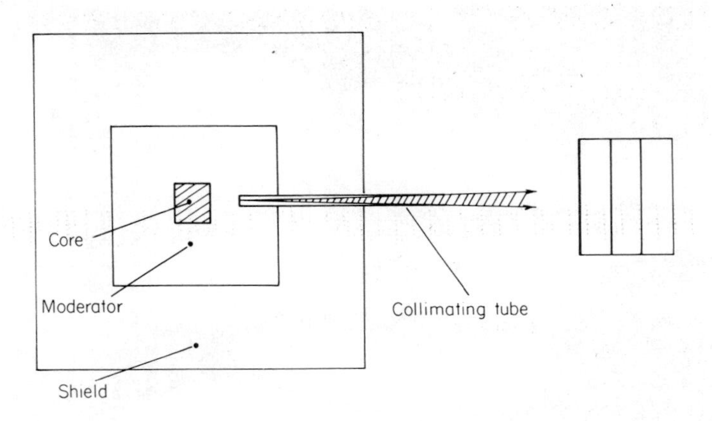

Figure 3 The extraction of a beam of approximately uni-directional neutrons from a reactor by means of a collimator. At the right hand side of the diagram is a view of the aperture of the collimator, measuring, say, 3cm x 2 cm and divided by vertical steel slats to give a beam of reduced angular divergence in the horizontal plane.

Figure 4 An instrument for recording diffraction patterns from powdered or polycrystalline material, installed at the PLUTO reactor at the Atomic Energy Research Establishment, Harwell. The large counter rotates in a horizontal plane, about a vertical axis on which is mounted the sample under observation . The sample can be mounted in a furnace or a cryostat as appropriate. (by courtesy of the U.K.A.E.A.)

relatively simple if all the incident neutrons have the same velocity and wavelength, or nearly so, and if they all strike the sample from the same direction. In practice it is only possible to attain extreme precision in uniformity of direction and of wavelength at the expense of a large reduction in the intensity of the beam. Accordingly a compromise is arranged whereby the directional spread of the incident beam is about $\pm \frac{1}{2}^{\circ}$ and the wavelength is constant to about $\pm$ 3%. A beam with an angular spread of direction of this amount can be achieved by channelling the neutrons to the face of the reactor by a collimator, indicated at C in Figure 3. The collimator will usually have a length of about a metre and an aperture of a few cm^2. It is often constructed of thin parallel plates of steel surrounded by material which has a high absorption for neutrons, yielding a beam of neutrons whose divergence in the horizontal plane is about $\frac{1}{2}^{\circ}$ but with a larger divergence, up to about 2°, in the vertical plane. The neutrons which emerge from the collimator, and are therefore travelling in approximately a single direction, have now to be 'monochromatised'. This is the name given to the process whereby a small fraction of them, of the order of 1% and whose velocities and wavelengths are near to a well-defined value, are preferentially selected:the remaining 99% are discarded. This selection is achieved by causing the stream of neutrons from the collimator to impinge at angle θ on a large single-crystal of a metal such as copper, lead or germanium. The reflected beam will be of approximately a single wavelength λ given by the Bragg equation $\lambda = 2d \sin\theta$ where d is the spacing of the atomic planes in the crystal which provide the effective reflecting surface. The value of θ will be chosen to suit the particular crystal employed, bearing in mind that a value of about 1 Å for the wavelength of the monochromatic beam is desired.

It has been realised from the very beginning of the development of neutron diffraction techniques that this process of monochromatisation, with an efficiency of only 1%, is an extremely wasteful way of handling the very expensive neutrons. As even more expensive reactors have been developed to provide more and more intense neutron beams, the need to invent more economical techniques has become increasingly apparent. Several possibilities are being tried and tested at the present time and we shall refer to the matter again later. Nevertheless, the simple, but wasteful, method using a monochromatised beam which we have outlined above is still used in the great majority of experiments and we shall assume that it is being employed in our subsequent discussion.

The monochromatic neutrons then fall on the sample under investigation which is mounted at the centre of a circular track which usually lies in a horizontal plane (as in Figure 4). The inclination of the sample to the horizontal and vertical planes can be varied very accurately within wide angular limits. A detecting counter, usually filled with boron trifluoride gas made from the ^{10}B isotope, rotates around the circular track in order to measure the rate of arrival of the scattered neutrons at any angular position. In most modern apparatus the motions of the sample and counter are programmed in advance in order that the surrounding space may be scanned automatically to measure and record the diffracted beams. The manner and extent to which the space is scanned will depend on the textural form in which the material under study is available, and this will be influenced very considerably by the nature and chemical constitution of the material itself. In principle the results of the investigation will be most precise and specific if the material is in the form of a single-crystal but it may often prove impossible for a sufficiently large specimen to be grown in this form: moreover, although the single-crystal method is the most powerful it is also the most time-consuming and a simpler investigation may quite often be adequate. The simpler, but much less informative method, is to examine the material in the form of a powder or a polycrystalline fragment - thus providing a neutron equivalent of the X-ray 'powder photograph'. In practice the neutron counter always rotates in a plane. With a powder sample the sample usually rotates about the same axis, perpendicular to this plane, to give a θ, a 2θ relationship, but with single-crystals it is generally necessary to scan the crystal in three dimensions. In order to increase the rate of data-collection multi-counter systems and position-sensitive detectors are being introduced.

It will be useful at this point to comment on the size of the sample, that is the amount of material which is required in order to allow an investigation to be made. This depends on two factors - first, the quantitative extent to which atoms scatter neutrons and, secondly, the intensity of the neutron beam which is available - and it is worthwhile to view these two factors alongside the more familiar case of X-ray diffraction. There is not a lot of difference between the scattering abilities of atoms for X-rays and neutrons - on average the X-ray values are greater by five or ten times - but the number of X-ray quanta which cross a unit area each second in an X-ray beam is about a thousand times greater than the number of neutrons in an average neutron beam. The consequence of this is that when a sample is to be studied with neutrons it needs to have a volume which is about a thousand times greater than would be sufficient for an X-ray study. In quantitative terms this means that a single-crystal for neutron study has to have a volume of about 8 mm^3 whilst in the form of powder about 500 mm^3 would be needed. These are average values which could be appropriate at research reactors which give a flux of about 10^{13} neutrons cm^{-2} sec^{-1}. There are reactors now operating in the U.S.A. and at Grenoble, France, which have fluxes of 10^{15} and these require correspondingly smaller samples, falling to about 0.5 mm^3 for a single crystal. Reactors with lower fluxes than 10^{13} are of only limited use for neutron diffraction studies, with the result that this work is concentrated in a small number of laboratories in the world.

It will now have become clear that it is not always possible to secure materials in single-crystal form which are sufficiently large for neutron studies, particularly in the case of alloy systems and compounds which undergo phase-transitions when their temperature is changed. As a result the experimenter often has to be content with polycrystalline material, especially for his preliminary investigations. Mention of phase-transitions and the influence of temperature makes this a convenient point to note that the neutron method is very adaptable to use over a wide range of temperature, from a few degrees to over 2000°K, and also at high pressures. The reason for this is that it is relatively simple, compared with the corresponding task with X-rays, to construct cryostats, furnaces and pressure vessels which are readily penetrated by the radiation. The absorption coefficients of solids for neutrons are so small that the loss of intensity by transmission through a few millimetres of constructional materials such as steel or aluminium is negligible. Accordingly, rigid enclosures in which the sample is subjected to changes of temperature and pressure, while immersed in the neutron stream, can readily be constructed. In this way samples can be observed under a variety of physical conditions and the experimental output, which usually consists of values of neutron intensity recorded on magnetic or paper tape in a programmed sequence, can incorporate also a record of temperature, pressure or magnetic field. Figures 4 and 5 show examples of diffractometers used for poly-crystal and single-crystal measurements respectively at the Atomic Energy Research Establishment, Harwell.

Recent increases in neutron intensity, leading to improved angular resolution in diffraction patterns, have greatly increased the amount of detailed information which can be obtained from a powder diffraction pattern in those cases where a suitable starting model of an atomic or magnetic structure is available for refinement. The method of powder-profile refinement, first suggested by Rietveld (1969)[1], depends on a least-squares refinement of all the intensity-ordinates in the pattern, at, say, intervals of 0.1° of the scattering angle, rather than a refinement of the structure factors normally obtained by integration over individual reflections in the pattern, if these are resolved. The success of the method rests on a prior knowledge of a model which has correctly identified those parameters which are to be refined.

With the foregoing account of some of the experimental details, aimed at giving a general impression of the technique from a practical point of view, we return to a study of the principles of the scattering of neutrons by atoms.

Neutron Scattering by Atoms

Although our final aim is to study the scattering of neutrons by magnetic materials it will be necessary for us, first of all, to look at the scattering by ordinary materials, particularly as we shall see in due course that it is necessary to identify separately the 'ordinary' scattering and the scattering which is magnetic in origin.

What we may consider as the 'ordinary' scattering arises from an interaction between the neutrons and the nuclei of the atoms which are being bombarded, and henceforth we shall call it the 'nuclear scattering'. It occurs for all atoms, whether they have magnetic properties or not, and simply indicates that it is the nuclei, wherein resides the whole of the mass of the atom, which act as the obstacle to the uncharged neutrons in their passage through a material. This is in sharp contrast to the situation for X-rays which are scattered only by the electrons which surround the nuclei. This happens because the X-rays, being a form of electromagnetic radiation, can set the electrons

Figure 5 A neutron diffractometer for the study of single-crystals. The detecting counter rotates in a horizontal plane and the crystal can be aligned in 3 - dimensions, thus ensuring that the whole of 3 - dimensional space, relative to the crystal axes, can be explored. (by courtesy of the U.K.A.E.A.)

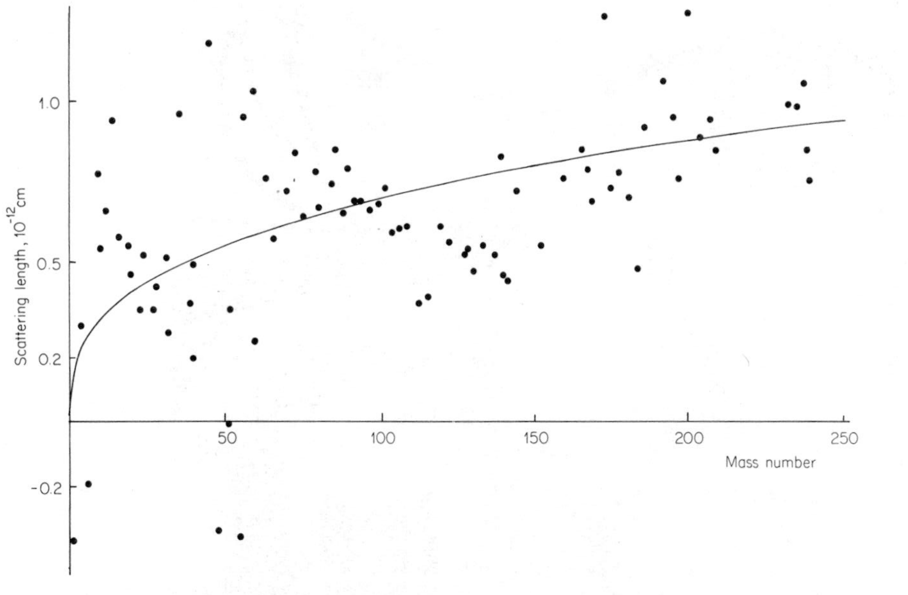

Figure 6 The variation of the neutron-scattering amplitude among the elements, shown in relation to the curve for potential scattering, whereby the scattering length is equal to $1.5\,A^{\frac{1}{3}} \times 10^{-3}$ cm, where A is the mass number. It will be noticed that divergences from this curve are especially large for light elements, the iron group of transition elements and the rare-earth elements.

into oscillation on account of their electrical charges and their very small mass.

It is of prime importance to examine the way in which the magnitude of the nuclear scattering varies among the different chemical elements, or, more strictly, among the different nuclides, where we recall that 'nuclide' is the name given to a specific nucleus with a defined value of both atomic number and mass number. It is found that the nuclear scattering depends on two features of the nuclide, first its size, or the amount of space which it occupies, and secondly, the details of its energy levels or, more strictly, the energy levels of the compound nucleus formed by combining a neutron with the initial target nucleus. Quantitatively, the amount of scattering can be expressed in terms of a "scattering length" b or a "scattering cross-section" σ. The former is defined by saying that the scattered wave which spreads out from a nucleus when it is bombarded by an incident neutron wave $e^{i\kappa z}$ of unit amplitude may be written

$$\psi = -\frac{b}{r} e^{i\kappa r} \qquad (1)$$

In this expression r is the distance from the nucleus at which the scattered wave is measured and κ is the wave-number of the neutrons. Thus, $\kappa = 2\pi/\lambda$ where λ is the wavelength of the neutron. The dimensions of b are those of a length and this is the reason why it is called the scattering <u>length</u>. A typical value of b is, as we shall see shortly, 0.5×10^{-12} cm. The "scattering cross-section" σ is defined by

$$\sigma = \frac{\text{outgoing current of scattered neutrons}}{\text{incident neutron flux}}$$

$$= \frac{4\pi r^2 v \left| \left(\frac{b}{r}\right) e^{i\kappa r} \right|^2}{v \left| e^{i\kappa z} \right|^2}$$

where v is the neutron velocity

This expression reduces to $\sigma = 4\pi b^2$, thus giving a simple correlation between cross-section and scattering length.

Of the two contributions to b and σ, which we mentioned earlier, that due to the nuclear size is usually called the 'potential scattering'. It can be shown to be equivalent in value to the nuclear radius, which, in turn, can be shown to increase as the cube root of the mass number A of the nucleus. This particular relationship is an expression of the fact that nuclei, being built up of neutrons and protons, are of constant density. Thus the nuclear radius R is given by

$$R = 1.5\ A^{\frac{1}{3}} \times 10^{-13}\ \text{cm} \qquad (2)$$

If the scattering were due to this 'potential scattering' alone, then b would increase slowly, but steadily, with the mass number. However the second component which we have mentioned, i.e. a resonance contribution which depends on the energy levels in the compound nucleus, results in seemingly random variations between one atomic species and the next as we advance through the table of atomic masses from light to heavier elements. The overall pattern is indicated in Figure 6 and a detailed study of the information which it contains will indicate some of the important applications of measurements of this nuclear scattering. Thus a light atom, such as hydrogen, is only inferior in scattering to the heavy atom lead by a factor of about two. Indeed its isotope deuterium attains a scattering length of about 70% of that of lead. By contrast, in the case of X-ray scattering the scattering length of lead is about 80 times as great as for either hydrogen or deuterium, simply because the lengths for X-rays are proportional to the numbers of electrons which the atoms contain. Another feature which is evident from Figure 6 is that there are often large differences in the neutron scattering lengths of neighbouring atoms in the Periodic Classification: for example, the values of b for Mn, Fe, Co, Ni are quite different and this makes it possible to distinguish these elements from one another quite readily and this is an achievement of immense value in attempts to interpret the structures and behaviour of their alloys. Finally, in considering Figure 6, we must emphasise that all the values of b which are shown there have been determined empirically by experiment. The need to use empirical values arises because, although the contribution to b from nuclear size can be easily calculated, insufficient is known about the details of nuclear structure to permit an accurate calculation to be made of the contribution from the resonance effect.

Magnetic Scattering

Let us now return to our main task of discussing magnetic scattering . What we usually call "magnetic materials" are basically of two types and they all contain atoms which carry a magnetic moment. In one class of material, which we call 'paramagnetic', the direction in which the magnetic moment of an individual atom points has no effect on the direction of pointing for its neighbours. As a result the moment directions, as we proceed through the material from atom to atom, are quite random. The second class of materials display 'co-operative' magnetism, which means that the direction of any moment influences the direction of its neighbours, with the result that the individual directions line up in ordered patterns. The simplest case of this second class of materials is the ferromagnet in which, within the volume of a domain containing many atoms, the directions of the magnetic moments are aligned in unison, all pointing in the same direction. However, the case of ferromagnetism is only the simplest, albeit the best known, case of co-operative magnetism. Antiferromagnetism, helimagnetism and other systems of linked moments are now known, largely as a result of investigations with neutrons, which we shall show to have the ability to discern both the magnitude and directions of the atomic moments.

The possession of a 'magnetic moment' by an atom or ion means that the atom or ion has a resultant electron spin. Each electron in an atom has a spin which may be clockwise or anticlockwise and as successive electrons are added, to build up heavier atoms, the tendency is for the electron spins to cancel out in positive-negative pairs. It is the atoms and ions in which this does not happen which show magnetic properties and which produce magnetic materials. For example, an Mn^{2+} ion in MnO has five unpaired electrons : an atom of iron in metallic iron appears, on average at least, to bear 2.2 unpaired electrons. It is these unpaired electrons, giving resultant spins, which cause additional scattering of the neutrons, additional that is to the basic scattering caused by the nucleus.

The reason for this additional scattering is that the neutron itself possesses a magnetic moment and it is the interaction between the moments of the neutron and the atom which yields the extra scattering. The reader may be puzzled to understand why a neutron, which possesses no net charge and no electric dipole moment can have a magnetic moment. This matter is not completely understood but an acceptable theory assumes that the neutron spends part of its time dissociated into a proton and a π meson

$$n \rightleftarrows p^+ + \pi^-$$

Although the centres of gravity of the spinning clouds of positive and negative charge do coincide, thus giving a zero dipole moment, it is postulated that one cloud is more diffuse than the other, thus giving a net value of magnetic moment.

We may then interpret the interaction between the moment of the neutron and that of the atom in terms of an additional scattering length p, which we term the 'magnetic scattering length'. We shall in due course have to consider whether there is any correlation between the nuclear and magnetic scattering and, in particular, whether the two processes of scattering can be considered completely independently of one another. Initially however we shall discuss the physical quantities which determine the magnitude of p. It can be shown that

$$p = \mu_0{}^2 \frac{e^2\gamma}{m} .Sf \qquad \left[p = \frac{e^2\gamma}{mc^2} . Sf \right]_{emu} \tag{3}$$

where γ is the magnetic moment of the neutron expressed in nuclear magnetons, i.e. the number 1.91; e, m are the charge and mass of the electron ; c is the velocity of light; S is the spin quantum number of the atom or ion and f is a form factor. It will be noted that the factor e^2/mc^2, which occurs in this expression, is the classical radius of the electron. Thus p, the magnetic scattering length, is of the order of the electron radius. If we substitute in this equation the quantitative values of the constants, then the equation becomes

$$\mathbf{p} = 0.54\ Sf \times 10^{-12}\ \text{cm} \qquad (4)$$

The form factor takes into account the finite size of the atom, from the point of view of the magnetic scattering. The magnetic electrons are distributed over a region of space which has dimensions of about 1Å (0.1 nm). This is about the same size as the neutron wavelength, with the result that the differences of phase between the scattered contributions from different parts of the atom become significant as the angle of scattering, 2θ, increases. The contribution from the atom as a whole falls off from its initial maximum value at $\theta = 0^{\circ}$ to a value which is only about 25% of this when $(\sin\theta)/\lambda$ has risen to 0.3 Å^{-1}. In contrast therefore to the angular invariance of the nuclear scattering, the magnetic scattering will become negligible at large angles. At small angles, however, the two scattering amplitudes are often of about the same magnitude. For example, for the ion Fe^{2+} the value of S is 2 and therefore, from equation (4), p is equal to 1.08×10^{-12} cm which is a little larger than 0.96, the nuclear scattering length for iron. For Mn^{2+}, for which S is equal to 5/2, the magnetic scattering amplitude is 1.35×10^{-12}cm and this is several times the nuclear scattering amplitude of manganese which happens to be rather below the average value for the elements. Figure 7 compares the form-factors of a number of elements and shows in particular how the form-factor falls off much more quickly with angle for Pd, Rh and Mo than it does for iron. This indicates that the spin density of the 4d electrons in the former group of elements is much more spread out in the space than is that of the 3d electrons which are relevant in the elements of the iron group.

The equation above, in which p is proportional to the spin quantum number, actually refers only to the simplest cases of magnetic atoms and ions, namely those in which the magnetic moment is due only to that part of the angular momentum which arises from the intrinsic electron spin. This is by no means an uncommon case, particularly among the elements of the iron group of transition elements, but in the general case there is also a contribution to the angular momentum which comes from the orbital motion of the electrons. In the iron group it is the 3d electrons forming the outermost shell of the atoms which give rise to the magnetic moments and they are affected by the field of the neighbouring atoms in such a way that their orbital momentum is rendered inoperative, or 'quenched'. On the other hand, for the rare-earth elements the magnetic properties are due to the 4f electrons in the N-shell and these, in this case, are shielded from the effect of neighbouring atoms by the intervening O shell. The orbital momentum is thus immune from the quenching action and equation (4) is replaced by

$$p = \mu_0 2 \frac{e^2\gamma}{2m} \cdot gJf \qquad \left[p = \frac{e^2\gamma}{2mc^2} gJf\right]_{emu} \tag{5}$$

where g is the Lande splitting factor. In this expression the effective form-factor f is compounded from separate values f_L, f_S for the orbital and spin moments. These two factors are weighted in proportion to the relative contributions which the two types of moment make to the resultant moment. Accordingly equation (5) leads to

$$p = 0.54\ (S_J f_S + \tfrac{1}{2} L_J f_L) \tag{6}$$

where the numerical values of S_J, L_J are

$$\{J(J + 1) \pm [S(S + 1) - L(L + 1)]\} \quad / \ 2(J + 1)$$

if the alternative + or - sign is taken in the two cases respectively.

Figure 8 shows some data for the rare-earth ion Nd^{3+} and compares the resultant curves, which include both spin and orbital contributions, as measured experimentally and calculated theoretically. It will be noticed that the measured form-factor curve falls off much more quickly with increasing value of θ than the calculated curve does, suggesting that the ions in the crystal are rather larger than a free ion would be. The figure also shows the two separate, and substantially different, contributions from the spin and orbital momenta.

Our discussion so far has concentrated on the values of the magnetic scattering amplitude p and the way in which this depends on the electronic constitution. In paramagnetic materials, where the direction of the magnetic moment varies at random from atom to atom, the <u>magnitude</u> of the moment, and hence of p, is our only concern and

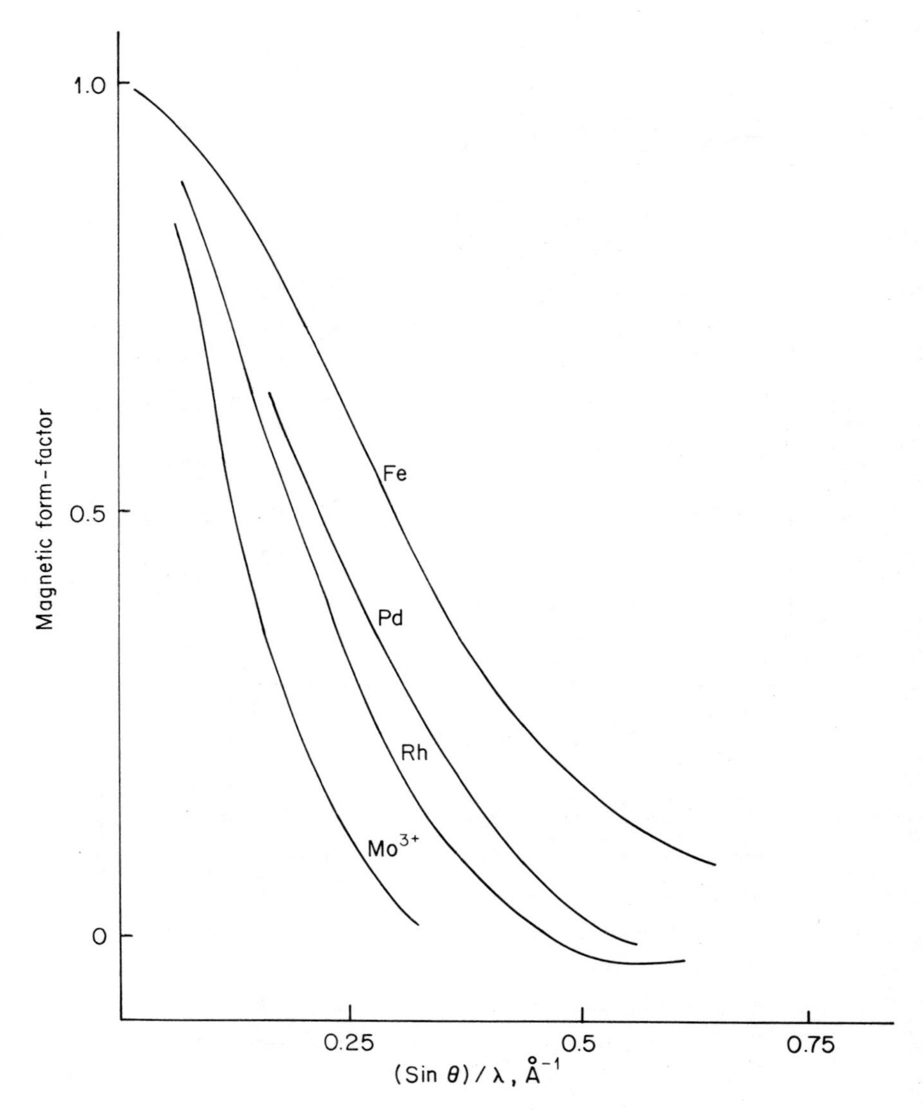

Figure 7 The variation with $(\sin\theta)/\lambda$ of the magnetic form factor of the metals Fe, Rh and Zr and for the Mo^{3+} ion.

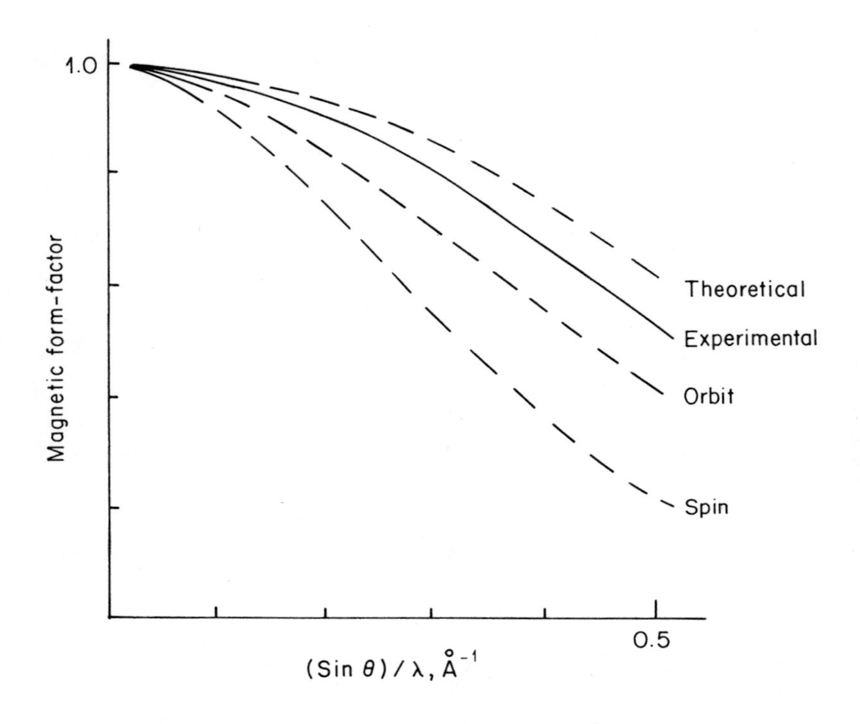

Figure 8 Comparison of the form-factor for Nd^{3+}, measured experimentally by Koehler and Wollan, with the theoretical calculation of Blume, Freeman and Watson (1962)[2]. The ground state of this ion is $4J_{3/2}$ with J = 9/2, S = 3/2, L = 6. Thus the spin and and orbital moments are antiparallel, so that the expressions for the spin and orbital parts of the form-factor (which are also drawn in the figure) appear with opposite signs in the expression for the resultant form-factor.

its direction is not of any significance. For co-operative magnetic materials, on the other hand, the direction is significant, and of extreme importance, and in these cases the direction of the moment has immense influence on the magnetic scattering of the neutrons. In fact the extent of the scattering is determined not simply by the value of p but also by the geometry of the scattering process: the important features and parameters can be discussed in terms of figure 9. Normally we shall be using a beam of unpolarised neutrons i.e. one in which the direction of the spins of the neutrons is random, and we shall consider the behaviour of such a beam first. In simple terms we can say that the effective value of p, given in equations (3), (5), is reduced by a factor $|\underline{q}|$ where the vector $\underline{q}$ is called the magnetic interaction vector which, in terms of figure 9, is defined by

$$\underline{q} = \underline{\varepsilon}(\underline{\varepsilon}.\underline{K}) - \underline{K} \qquad (7)$$

Here $\underline{\varepsilon}$ is a unit vector in the direction of the normal to the scattering plane, and $\underline{K}$ is a unit vector in the direction of the magnetic moment of the scattering atom. It can be shown from this definition that the magnitude of the vector $\underline{q}$ is equal to sin β and it lies in the plane of $\underline{\varepsilon}, \underline{K}$ in a direction which is normal to $\underline{K}$. the significance of this dependence of the scattering on the product $|\underline{q}|$ p is that an experimental study of the distribution of the magnetic scattering in the diffraction patterns will enable us to determine the directions of the magnetic moments of the individual atoms, relative to the axes of the crystallographic unit cell of the material. In this way the neutron method enables us to identify magnetism on an atomic scale, detailing the magnitude and direction of the magnetic moments of the individual atoms within the unit cell which builds up the structure.

By adding up the magnetic contributions from the individual atoms in the *cell* we arrive at the overall amplitude for any reflected beam, so far as magnetic scattering is concerned, giving a structure factor $F_{magn.}$ Quite separately, we could add up the nuclear scattering contributions for the atoms in the unit, giving $F_{nucl.}$ How do we combine these two types of scattering? The answer is that, for unpolarised neutrons, the

<u>intensities</u>, and not the amplitudes, are additive. This means that the resultant intensity of the hkl reflection is proportional to $|F_{hkl}|^2$, where

$$|F_{hkl}|^2 = |\sum_n b_n \exp 2\pi i(hx_n + ky_n + lz_n)|^2 \quad (8)$$

$$+ |\sum_n \underline{q}_n p_n \exp 2\pi i (hx_n + ky_n + lz_n)|^2$$

In this expression $x_n, y_n z_n$ are the fractional co-ordinates of the nth atom in the unit cell.

If the neutrons in the incident beam have their magnetic moments in some defined direction, i.e. if the beam is polarised with a polarisation vector indicated by λ in figure 9, then the behaviour is more complicated. There will be interference between the nuclear scattering amplitude and the magnetic scattering amplitude and it is no longer permissible simply to add together the two intensities. The result is a structure amplitude factor for the unit cell.

$$F_{hkl} = \sum_n [b_n + (\underline{\lambda}.\underline{q})p_n] \exp 2\pi i(hx_n + ky_n + lz_n) \quad (9)$$

We can correlate this with our previous expression as follows. If our incident beam is unpolarised then we have to write a term of the above type for each direction of polarisation. There will be no phase - correlation between the different directions and we can assess the resultant intensity by adding together the values of $|F_{hkl}|^2$ for the individual components. When this addition is done the cross term $\underline{\lambda}.\underline{q}$ will average to zero, thus leading to the expression which we have already quoted as equation (8).

A more complete treatment of the effect of polarisation acknowledges that not only does the scattering amplitude depend on polarisation but, also, there may be changes in polarisation as a result of the scattering process. The scattering amplitude is described by

$$U = (u'd') \{ b + p\underline{q}.\hat{\underline{\sigma}} \} \binom{u}{d} \quad (10)$$

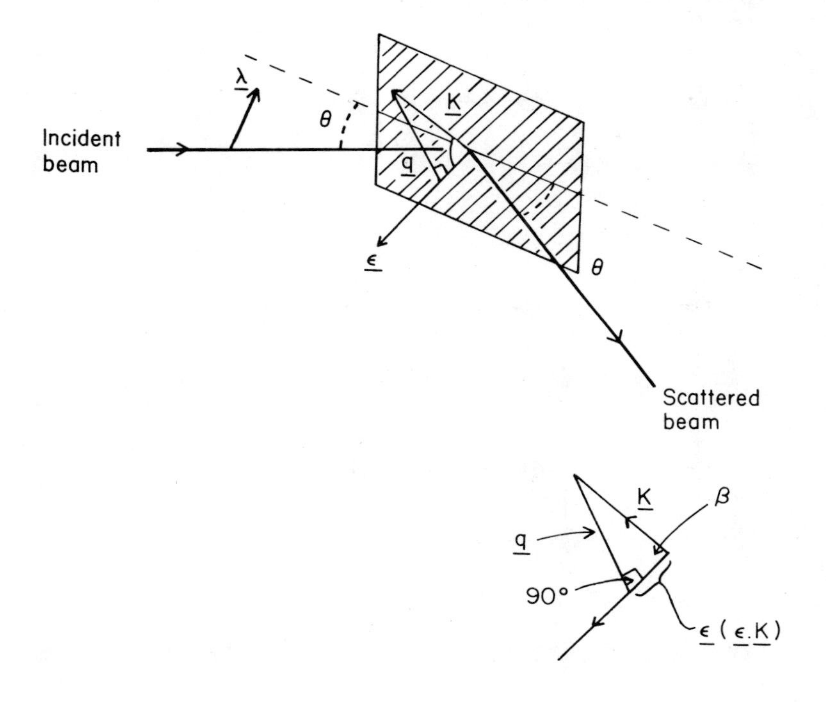

Figure 9 A diagram to indicate the parameters $\underline{q}$, $\underline{\varepsilon}$, $\underline{K}$, β and $\underline{\lambda}$ which determine the magnitude of the magnetic neutron scattering and the way in which this is influenced by the orientation of the magnetic moments. The lower portion of the diagram illustrates the truth of the expression for $\underline{q}$ given by equation 7 of the text.

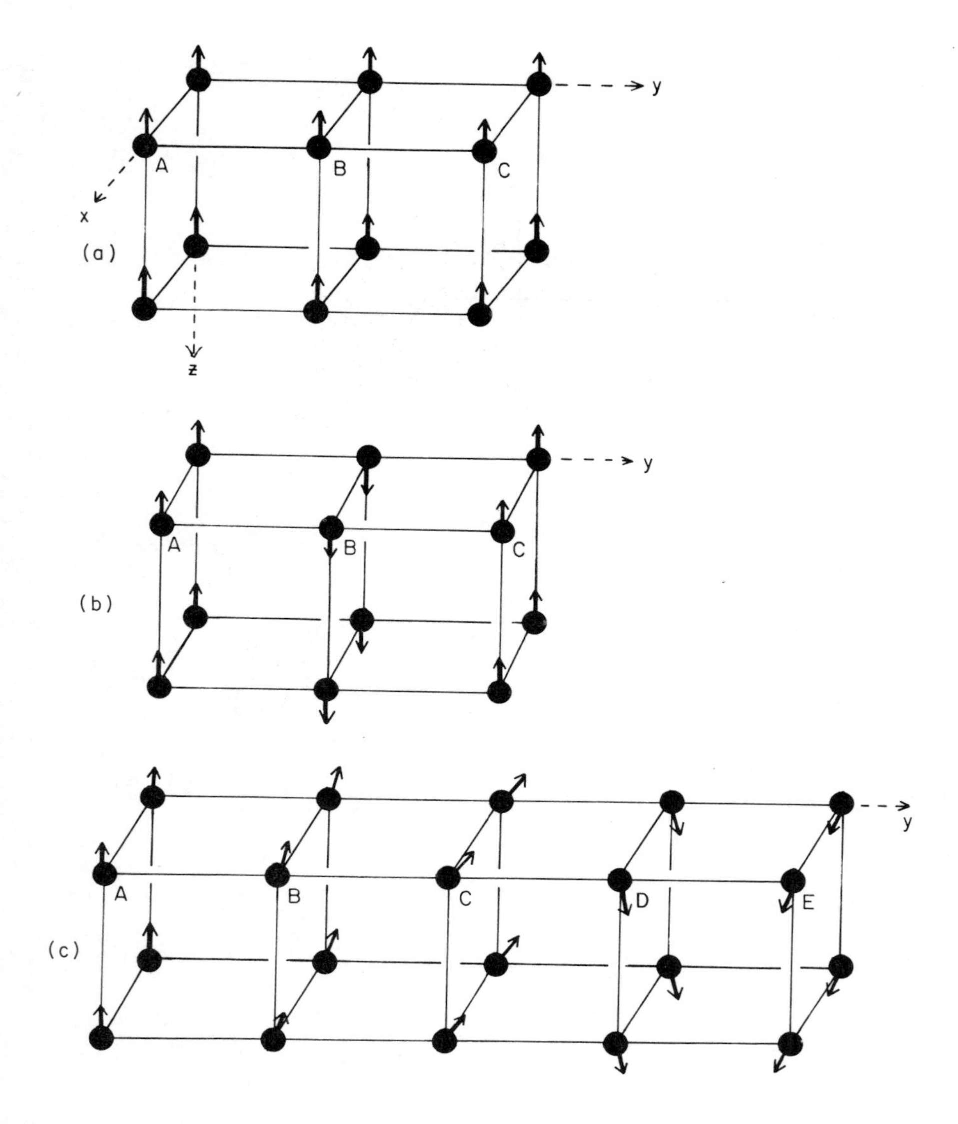

Figure 10 Diagrammatic representation of different types of co-operative magnetic arrangements: (a) ferromagnetic (b) antiferromagnetic and (c) helimagnetic. The arrows on the individual atoms indicate the orientation of their magnetic moments.

where $\hat{\underline{\sigma}}$ is the Pauli spin operator and the matrix elements u,d and u', d' refer to the spin states of the incident and scattered beams as 'up' or 'down'. It can be shown (see Moon, Riste and Koehler, 1969)[3] that this expression leads to four separate amplitudes which determine the numbers of neutrons whose spins have changed, or remained unchanged, in the scattering process. A technique of "polarisation analysis" is used to assess these four amplitudes separately and leads to direct methods of separating the nuclear and magnetic scattering and, incidentally, of distinguishing between coherent and spin-incoherent scattering. This technique has immense possibilities, but, depending as it does on the availability of very intense beams of neutrons, it has not yet been very widely applied.

Description and Evaluation of Magnetic Structures

Consideration of equation (8) will indicate the fundamental differences which are to be expected in the diffraction patterns given by materials having different types of co-operative magnetic arrangement. Let us consider the three different magnetic arrangements illustrated in figure 10. Each of these is constructed of identical atoms bearing a magnetic moment whose direction is indicated for the various atomic sites by the superimposed arrows. First of all we note that if diffraction patterns were taken of each of these structures with X-rays then the patterns would be identical, because X-rays are not able to recognise the existence of the magnetic moments. We can illustrate this X-ray pattern by curve (i) in figure 11. With neutrons , however, the patterns from the three structures will be quite different from one another. For structure (a) which is ferromagnetic the pattern can be represented by curve (ii) in figure 11. It will be noticed that the angular positions of the lines in this curve are exactly the same as for the X-ray pattern (i)*, simply because the repeating unit for structure (a) is the cube of side AB, exactly the same as if no arrows were marked on the structure. On the other hand the relative intensities of the lines in curves (i), (ii) are different. Further investigation would show that the temperature dependence of the two patterns, also, was different and that the lines

* This assumes, of course , that the X-rays and neutrons have the same wavelength.

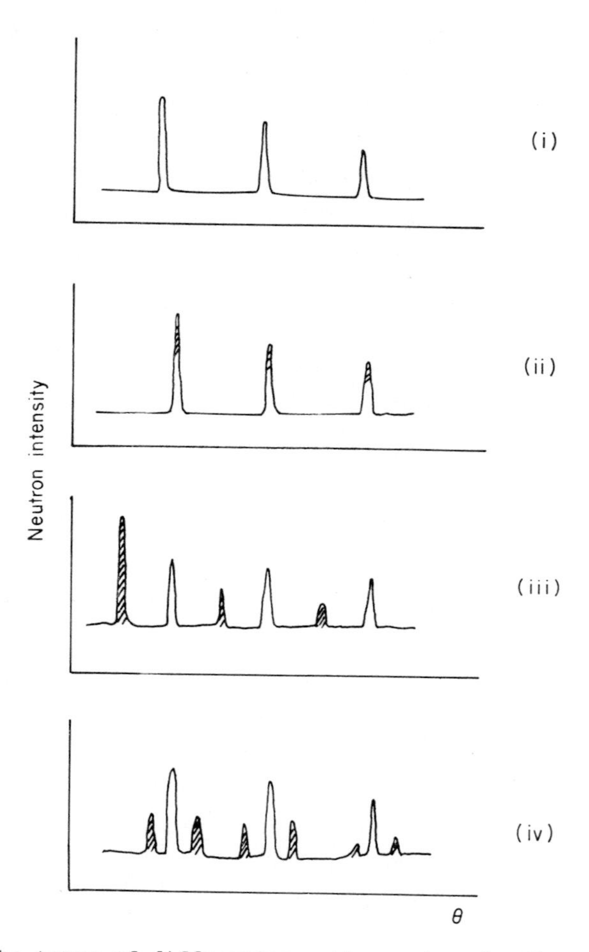

Figure 11 The types of diffraction pattern given by the various magnetic structures which were illustrated in Figure 10. Curve (i) is typical of X-ray diffraction patterns for any of the structures 10(a), 10(b) and 10(c): Curve (ii) is a neutron pattern for structure 10(a): Curve (iii) is a neutron pattern for structure 10(b) and Curve (iv) is a neutron pattern for structure 10(c). In each of the neutron patterns the shaded areas represent the contributions to the scattering which are of magnetic origin.

in curve (ii) could be sub-divided into two parts, which are shown with different shading, and which are affected differently when the temperature is altered. In fact the separately shaded contributions are due respectively, to the nuclear and the magnetic scattering of the neutrons. The former, like the X-ray intensities, will vary only slowly and steadily with temperature under the influence of the Debye-Waller factor. The magnetic scattering will reduce rapidly as the ferromagnetic Curie temperature is approached and will be zero at temperatures higher than this. Moreover, the size of the shaded portion, in relation to the unshaded nuclear scattering, becomes steadily smaller as the value of sin θ increases. This is because of the angular variation of the form-factor for magnetic scattering which, as we have seen, falls off rapidly in value as the value of (sin θ)/ λ increases.

Turning to structure (b) in Figure 10 which is antiferromagnetic, we shall find that the neutron pattern, indicated at curve (iii) in Figure 11, is quite different. Its distinctive feature is that additional 'lines' are present and when these are indexed in terms of the normal unit cell, of side AB, it is found that half-integer numbers appear among the indices. For example, the indices of line P turn out to be 0,½,0, which at first-sight seems incompatible with the rule that planes of reflection have to correspond to whole-number values of Miller indices. The explanation is that the side of the unit cell, when the magnetic moments are taken into account, is twice as large in the y-direction as the length AB: the side of the unit in this direction is AC. If we had indexed the pattern (iii) in terms of a unit cell whose y dimension was AC (= 2 AB) then the indices of reflection P would have been 010, comprising only integral numbers. We see therefore that the appearance of the extra line P, and also Q,R, is direct evidence for the doubled unit-cell which is the consequence of the antiferromagnetism. If we now raise the temperature of the material whose structure is represented by diagram 10 (b) we shall find that the intensities of lines P,Q,R in 11(iii) fall off rapidly as the Néel temperature is approached. Beyond this temperature at which the ordered magnetic arrangement disappears, the lines P,Q,R are absent.

Structure (c), and the diffraction pattern which it gives (Figure 11(iv)), introduce quite new features. Such a material is described as being 'helimagnetic' from the helical manner in which its magnetic moments are arranged. The magnetic moments lie in sheets, perpendicular to the axis Oy in Figure 10(c), and within an individual sheet the moments all point in the same direction. However the orientation of this direction varies from sheet to sheet, rotating around by a constant angle - equal to about 55° in the figure - as we advance from plane to plane along the direction Oy. The effect of this on the diffraction pattern for neutrons is that each of the ordinary reflected beams is accompanied by a pair of satellites, one at a smaller angle θ than the ordinary line and the other at a higher angle, as illustrated in curve (iv)of Figure 11. From the precise positions and intensities of these satellite lines it is possible to deduce the details of the magnetic structure. The characteristic grouping of the satellites gives the first clue from an experimental diffraction pattern that it is produced by a helimagnetic structure. The next indication comes when the lines are indexed, in terms of the normal unit cell which would be known from observations using X-ray diffraction. The indices which are deduced will not all be integers and the non-integral values may not even be simple fractions. If we represent a normal reflection by a reciprocal lattice vector $\underline{k}$ corresponding to a reciprocal lattice point hkl, then the satellite reflections will be found at two points $\underline{k} \pm \underline{g}_o$ in reciprocal space. The vector $\underline{g}_o$ is directed parallel to the direction in space about which the moment-vector directions rotate, i.e. OY in figure 10, and the value of g_o is given by

$$|g_o| = \frac{\phi}{2\pi} \cdot \frac{a}{x} \tag{11}$$

where ϕ is the angle through which the moment-direction rotates as we advance from sheet to sheet; x is the distance between successive sheets and a is the distance between corresponding points in successive unit cells as we go along the spiral axis. In the case of the structure which we illustrated in Figure 10, $x/a = \frac{1}{2}$ and $\varphi/2\pi$ is approximately 0.15 so that $g_o = 0.3$. As an example, we may consider the structure in Figure 10 to represent the alloy Au_2Mn, which has tetragonal symmetry and for which the spiral axis is the [001] direction. The value of the vector $\underline{g}_o$ in reciprocal space is therefore 0,0,0.3. Accordingly, the 002 reflection in the diffraction pattern is accompanied by two satellites which index as 0,0,1.7 and 0,0,2.3 : in a similar way the reflection 101 would have satellites 1,0,0.7 and 1,0,1.3. On the other hand, in a powder diffraction pattern the reflection 110 would be accompanied by only a single satellite indexing as 1,1,0.3 because the reflections 1,1, + 0.3 and 1,1, - 0.3 have the same interplanar spacing and occur at the same position in the powder diffraction pattern.

At the same time, we can determine the intensities of the satellite reflections by evaluating the second term in equation (8), which is the contribution from magnetic scattering to any reflection (at the positions of the satellites there will be no contribution from nuclear scattering).
This gives the expression

$$F_{magn} = -\frac{e^2\gamma}{mc^2} \cdot \sum_n \left[\underline{S}_n - \underline{\varepsilon}(\underline{\varepsilon}.\underline{S}_n)\right] f_n \exp 2\pi i(hx_n + ky_n + lz_n) \qquad (12)$$

for the magnetic structure factor, where we have replaced $\underline{K}\,S_n$ by its equivalent $\underline{S}_n$ which expresses the magnetic spin as a vector. For the spiral structure in Figure 10 this leads to

$$|F_{magn}|^2 = \frac{1+\cos^2\phi}{4} \left|\sum_n p \exp 2\pi i(hx_n+ky_n+l'z_n)\right|^2 \qquad (13)$$

where $l' = l + g_o$ represents the satellite position and ϕ is the angle between the normal to the hkl plane and the spiral axis.

In this way we see in outline how the contribution from magnetic scattering can be calculated, whether to 'ordinary' reflections for a ferromagnetic, or to extra reflections for an antiferromagnetic or to satellite reflections for a helimagnetic material. The converse process, which is the concern of the experimenter faced with the problem of deducing the details of an unknown structure from a measured diffraction pattern, is more difficult, but by trial-and-error based on experience it can often be achieved, though it will not always be possible to achieve an unambiguous solution. This possible ambiguity is one of the limitations set by the use of polycrystalline samples or , at the best, single-crystals which do not correspond to single-domains. If there is to be no ambiguity of interpretation of the experimental data, then it is necessary that the structures in figure 10 should be maintained as single magnetic domains throughout the whole volume of a single-crystal. In certain circumstances, by the application of magnetic fields or strains, this can be achieved, but usually a single-crystal will consist of many domains. Thus for the structure (b) in figure 10 a single-crystal would usually be constituted for a series of separate domains for which the unique magnetic axis was not necessarily Oy, as drawn, but could also be Ox or Oz instead. Indeed in most cases only polycrystalline material is available, particularly for alloy systems and materials where a number of phase changes occur as the temperature is varied, so that the interpretation has to be indirect. If a correct model can be postulated then it may be refined accurately by profile-refinement, as we described earlier. A further disadvantage is that it is not very often possible to use direct methods of symmetry determination for magnetic structures, in contrast to the straight-forward use of space-group theory in arriving at the internal structure of an ordinary, non-magnetic, material. A first approach to a formal theory of magnetic symmetry is provided by adding to the normal symmetry operations a new operation R which reverses the direction of a magnetic moment. In this way Russian crystallographers have extended the concept of the normal 230 space groups to the 1651 so-called 'Shubnikov groups'. Of these, 1421 are applicable to the structures of ferromagnetic and antiferromagnetic materials. In practice it is not usually possible to isolate sufficient individual magnetic reflections to utilise the formal knowledge of the magnetic groups which has been built up in recent years. We have to bear in mind that the form factor f limits the intensities of reflections at high angles and also that the anisotropy of the unit cell is often very small in materials where cubic symmetry is only slightly distorted towards tetragonal or rhombohedral arrangements, with the result that individual reflections may be unresolved. Moreover the Shubnikov treatment of symmetry is not applicable to the helical spin structures where successive spins are rotated through an arbitrary angle of ϕ degrees, as distinct from angles of 180°, 90° or 60°.

Many variants and combinations of the magnetic structures which we have already described have also been reported. The most common is the ferrimagnetic arrangement found in the ferrites, where the magnetic spins on particular sites are oriented +, - in opposite directions but the magnitudes of the two spins are not the same. Such substances, of which magnetite Fe_3O_4 shown in Figure 12 is the simplest example, do show a resultant magnetic moment - unlike antiferromagnetic materials - but the moment is much smaller than it would be if all the spins were aligned up in the same direction, as they would be in a ferromagnetic. With magnetite itself it has also been found possible with neutrons to explain the phase change which is known to occur at $119^{\circ}K$ from thermal, magnetic and electrical measurements. The details of the magnetic scattering, when measured below this temperature, show that ordering takes place amongst the Fe^{2+}, Fe^{3+} ions which are on the octahedral sites i.e. among the dotted/shaded atoms in Figure 12. In the section of the unit cell which is shown in this figure the upper pair of these atoms become Fe^{3+} and the lower pair are Fe^{2+}. A further example of how distinctions may be made among the atoms of a given element is provided by the alloy Fe_3Al. The structure of this alloy is based on a body-centred cube and has long been known from X-ray work to include three different types of atomic site, two of them occupied by iron atoms and the third by aluminium atoms. It has been shown from measurements of the intensities of the neutron diffraction lines that the magnetic moments on the two different types of iron site are substantially different. The moment on the so-called A sites, which have 4 aluminium and 4 iron atoms as neighbours, are $1.46 \pm 0.1\mu_B$ whereas the D sites, which have 8 iron atoms as nearest neighbours, bear a moment of $2.14 \pm 0.1\mu_B$.

Other materials have been found to possess what can be considered as a ferromagnetic spiral, produced by combining the simple spiral of Figure 10(c) with a ferromagnetic arrangement: in this case the spin directions are not at right angles to the axis of the spiral. An example of such a structure is provided by the low-temperature form

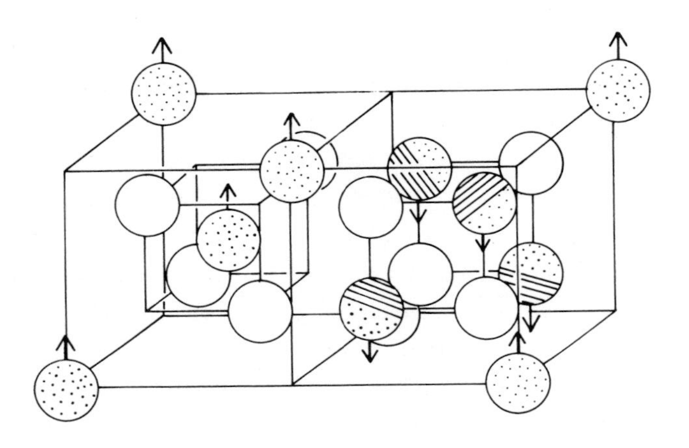

Figure 12 The structure of magnetite, Fe_3O_4, the best-known ferrimagnetic material, showing one-quarter of the complete unit cell. The dotted atoms are Fe^{3+}, of which there is one for each formula-unit, with an upward-pointing magnetic moment of 5 μ_B. The dotted/shaded atoms indicate Fe^{2+}, Fe^{3+} in equal numbers, arranged at random among the sites indicated: there is one of each ion per formula unit with downward-pointing moments of 4 μ_B, 5 μ_B respectively. Accordingly the net moment will be 4 μ_B, pointing downwards, per formula unit. The plain circles represent oxygen atoms.

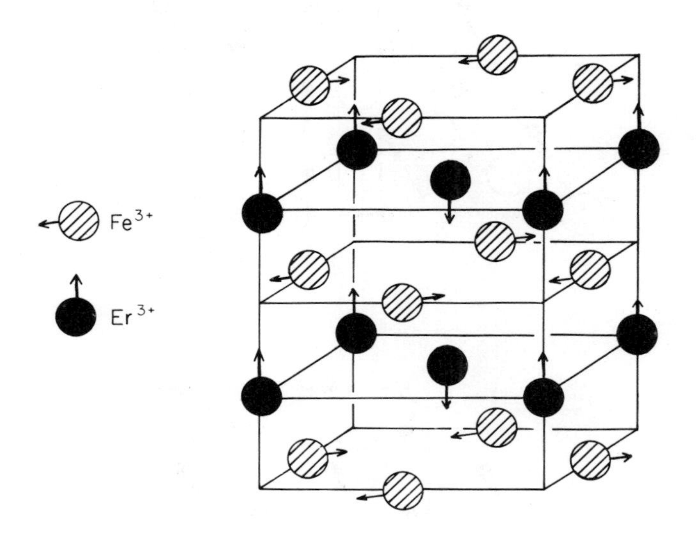

Figure 13 The magnetic structure of $ErFeO_3$ below 4.3°K. The moments of the ferric ions lie in horizontal planes, forming an antiferromagnetic arrangement. The moments of the erbium ions are directed vertically, parallel to the c-axis: each vertical line of downward pointing moments is surrounded by four vertical lines of upward pointing moments, and vice versa. Above 4.3°K the order among the erbium ions breaks down but the regular arrangement among the iron ions persists up to 620°K. (Koehler, Wollan and Wilkinson, 1960)[5]

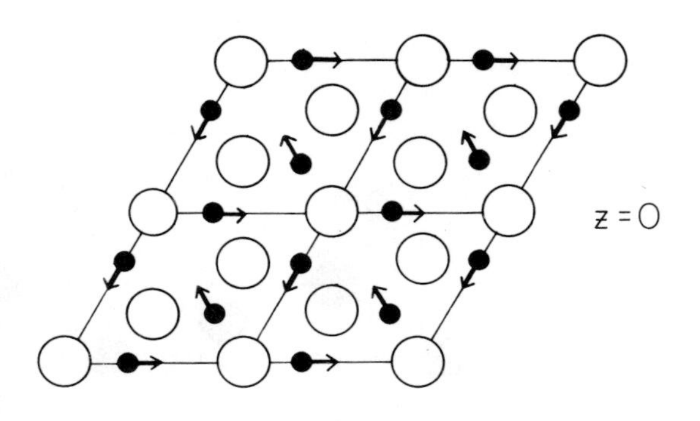

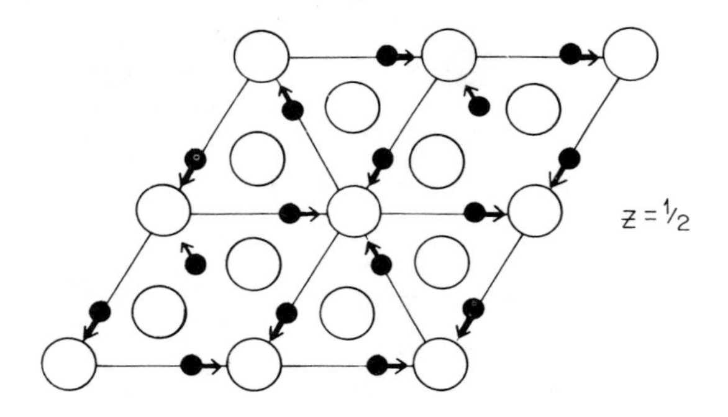

Figure 14 The magnetic ordering of successive layers of manganese ions (black) in the compound $YMnO_3$. The larger plain circles indicate oxygen atoms. The yttrium atoms , together with more atoms of oxygen, are sandwiched between the layers which hold the manganese ions.

of holmium, which displays a combination of a spiral component of 9.5 μ_B perpendicular to the c-axis with a ferromagnetic contribution of 2.1μ_B parallel to this axis.

The rare-earth elements and their alloys are remarkable for the intricacies of magnetic structure which they display and for the manner in which they undergo magnetic transformations at low temperatures. Thus erbium at very low temperatures displays a conical spiral structure which transforms at 20K to an antiferromagnetic arrangement combining a spiral with antiphase domains. At 52K there is a further change to a structure which has modulated moments directed along the c-axis , prior to a change to paramagnetism at 85K. A very useful summary of the varied magnetic structures which are found among the rare-earths has been given by Koehler (1972)[4]. Some of the perovskite structures are also worthy of mention, particularly those such as $HoFeO_3$ and $ErFeO_3$ which contain two very different kinds of magnetic atom i.e. the rare-earth element and iron. It these cases it is found that at ordinary temperatures only the iron atoms are magnetically ordered, giving antiferromagnetic structures, but when cooled to very low temperatures (4.3K in the case of $ErFeO_3$) ordering begins to occur among the moments of the rare-earth ions which were previously directed randomly in space. The resulting structure is illustrated in Figure 13.

Among other varieties of non-collinear arrangements of moments are the triangular arrangements displayed, for example, by rare-earth - iron- group oxides such as $YMnO_3$. This has a hexagonal layer structure with a unit-cell built up of six layers of oxygen atoms. The manganese atoms lie within oxygen layers at $z = 0$ and $z = \frac{1}{2}$ but the yttrium atoms constitute separate layers at $z = \frac{1}{4}$ and $z = \frac{3}{4}$ which interleave layers of oxygen. Two possible arrangements have been suggested for the magnetic structure. One of these, the α model, is illustrated in Figure 14 which shows a triangular arrangement of moments, directed <u>along</u> the hexagonal axes. It will be noticed that corresponding atoms in the layers at $z = 0, \frac{1}{2}$ have identical directions for their moments. It can be shown that an alternative, β , structure in which corresponding atoms at $z = 0, \frac{1}{2}$ have oppositely-directed moments is equally acceptable in explaining the experimental data if the moment directions are chosen perpendicular to the hexagonal axes. Another well-known non-collinear structure is the "umbrella" arrangement found in chromium selenide and illustrated in Figure 15. The non-coplanar moments of the chromium atoms lie along the generators of a cone and are oppositely directed in successive layers of atoms.

It must again be emphasised that there is no direct way of deducing many of these structures from the experimental data. Knowledge of them, and their postulation in the first instance, has arisen by trial-and-error, and a great deal of ingenuity, when it has become evident that the observed diffraction patterns have not been consistent with models based on the simpler types of magnetic arrangement. In practice an experimental investigation of a magnetic structure is generally made when there is some evidence, from physical properties such as resistivity or magnetic susceptibility, that a phase change is occurring at or near some particular temperature. The neutron diffraction pattern is then examined, usually with powdered or polycrystalline material, over a range of temperature on either side of the expected transition. A survey of the reflections obtained and their temperature dependence, with particular reference to any reflections which are not observed using X-rays, will then give a primary indication of those contributions to the diffraction pattern which are of magnetic origin. The next stage is to attempt to correlate the magnetic intensities with quantitative calculations in terms of the commoner types of magnetic structure. Depending on the degree of success obtained in this way it may become necessary to consider more complicated models. In seeking quantitative agreement it will be necessary to assess, and allow for, any preferred orientation of crystallites in polycrystalline material and it will be of great advantage to make further measurements with single-crystal if these can be obtained.

Before leaving this topic of the practical task of determining a magnetic structure we shall comment on a procedure of great value with ferromagnetic materials in indicating which parts of a diffraction pattern are of magnetic origin. From our discussion of the magnetic interaction vector $\underline{q}$ it is evident that the value of q would be zero if, in Figure 9, a magnetic field is applied in a direction normal to the reflection plane. This means that for a ferromagnetic material we are

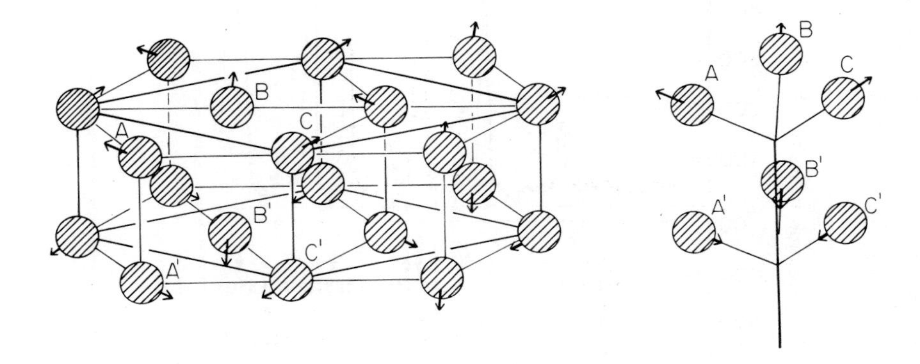

Figure 15 The non-colinear magnetic structure of chromium selenide, CrSe, showing the different directions, together with their reversals, in which the magnetic moments of the chromium atoms point. The magnetic cell is three times as large, in volume, as the chemical cell and the magnetic moment directions form the umbrella-type arrangement which is drawn separately on the right-hand side of the figure.

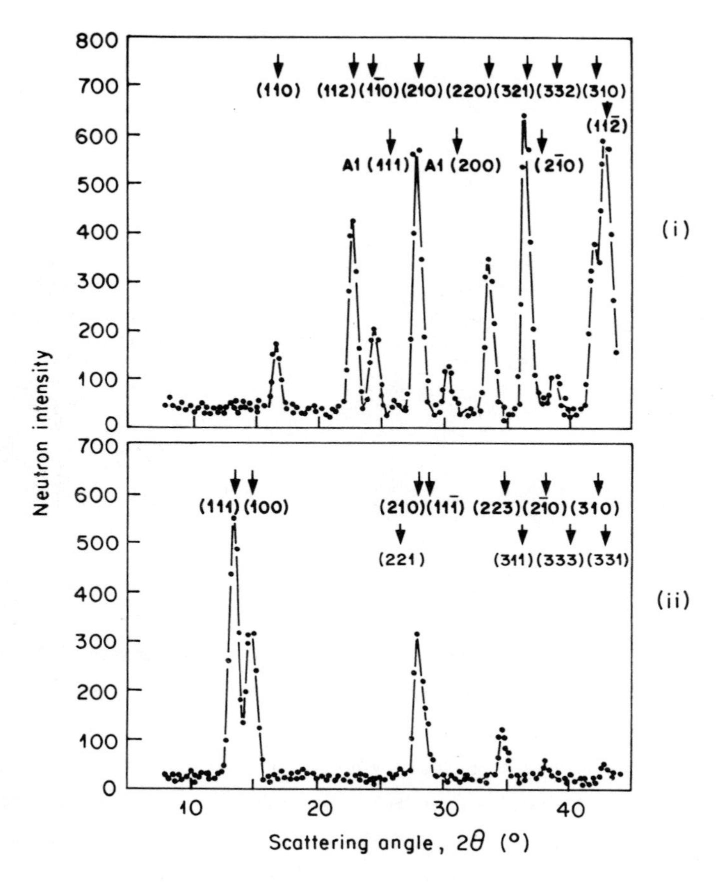

Figure 16 Powder diffraction patterns of α-Fe_2O_3 observed by polarisation analysis. In (i), where only neutrons of unchanged spin direction are recorded, the peaks are due to nuclear scattering only. In (ii) the neutrons have changed their spin direction and the peaks indicate magnetic scattering.
(Moon, Riste and Koehler, 1969)[3]

able to 'switch off' the magnetic scattering by applying a saturating field in this particular direction, and this procedure enables us to identify directly how much of the scattering is magnetic in origin. We also note that if we apply the field in a direction normal to ε, for example vertically in Figure 9, then the numerical value of q is unity and the amount of magnetic scattering is a maximum. If _no_ magnetic field is applied then the magnetic moments will usually be oriented at random in space in a polycrystalline sample or along one or possibly several easy directions of magnetisation in a single crystal. The average value of q^2 will then be an intermediate one. In the particular case of polycrystalline or powdered material the average value of q^2 is $\frac{2}{3}$ so that we can record, in turn, intensities proportional to b^2, $b^2 + \frac{2}{3}p^2$ and $b^2 + p^2$ by suitable choice of the field direction. This enables us to assess the magnetic scattering in absolute quantitative terms. For antiferromagnetic materials a distinction between the nuclear and magnetic scattering can be made very elegantly by polarisation analysis. An example is given in Figure 16 which shows two powder diffraction patterns of α-Fe_2O_3 using a polarised beam of neutrons for which the spin direction $\underline{\lambda}$ is along the scattering vector $\underline{\varepsilon}$. The upper diagram records only neutrons whose polarisation direction is unchanged in the scattering process, and, under the specified conditions of $\underline{\lambda}.\underline{\varepsilon} = 1$, it can be shown that these neutrons represent purely nuclear scattering: the lower diagram records neutrons whose spin has been reversed and here the peaks shown represent purely magnetic scattering. The use of polarisation analysis for making distinctions between different types of scattering, in this and similar ways, is restricted at present by the rather low efficiency of reflection of the crystals used for producing the polarised beam. The commonly used alloy $Co_{0.92}Fe_{0.08}$ is at a disadvantage because of the relatively high absorption coefficient of cobalt for neutrons. Development work with crystals of Heusler alloys for use as a polariser promises to be successful.

Non-Stoichiometric Alloys

Our discussions so far have related to elements, such as the metals iron (ferromagnetic) and chromium (antiferromagnetic), or to compounds of defined composition, such as the oxide and fluoride MnO and MnF_2, in which all the constituent atoms occupy well-defined positions in the ordinary chemical unit cell. We have also mentioned certain alloys such as the helimagnetic Au_2Mn, and in these cases, too, we were again concerned with a well-defined composition and chemical structure. There are however many alloy systems in which a given crystallographic structure may exist over a wide range of composition. In such cases the overall atomic proportions are not represented by simple integral numbers and it is inevitable that there are inhomogeneities in the local environment, simply because each atom necessarily has an integral and well-defined number of neighbours. As a result of the inhomogeneities the magnetic environment and the magnetic forces between an atom and its neighbours may vary significantly from point to point in a material. In these circumstances it has been observed that two magnetic structures may co-exist within certain ranges of composition of the alloy. An example is the platinum-iron system in the region of 20-30 per cent of iron (Bacon and Crangle, 1963)[6]. At the ideal composition of 25%, corresponding to Pt_3Fe, and for a few per cent of iron below this, there exists the antiferromagnetic structure illustrated at (a) in figure 17 in which there is an antiferromagnetic arrangement of sheets which are parallel to the (110) planes. When iron atoms are introduced beyond 25% a second structure , shown at (b) in the figure, appears, in which the sheets are parallel to the cube faces, and over the region of 25-30% of iron the two structures exist together. Beyond 30% of iron the first type of structure has almost disappeared. The magnetic diffraction lines are invariably sharp and it is concluded that in the intermediate region there is a phase-coherence over many unit cells, giving an inter-twining of the two structures, as distinct from segregarion into distinct domains of one or the other kind. X-ray diffraction photographs give no indication of the existence of two separate phases. In a similar way it has been found that antiferromagnetic and ferromagnetic structures can exist together. An example of this is provided (Bacon and Mason, 1967)[7] by the series of alloys which extends from Au_2Mn_2 to Au_2MnAl, as some of the manganese atoms in AuMn are replaced by aluminium. The former of these alloys is antiferromagnetic and the latter is ferromagnetic. As the atoms of aluminium are introduced the number of manganese-manganese nearest neighbours decreases and a ferromagnetic alignment is encouraged. In the intermediate region there will be statistical variations in the environment and here both structures are found to exist together.

Foreign atoms and dilute alloys

In the previous paragraph we have discussed the effect on magnetic structures of the inhomogeneities which are necessarily present in non-stoichiometric materials. An extreme example of this variability of environment is provided by the very dilute alloys where a relatively few 'foreign' atoms are introduced into a matrix material. If the alloy is very dilute the foreign atoms will be so far apart that they will exert a negligible effect on each other and we shall indeed have the opportunity of assessing the effect of the surrounding matrix on an individual atom. In a more thorough examination, our description of this situation will also include the effect of the impurity atom on the magnetic order in the surrounding material. Collins & Low (1965)[8] at Harwell have shown that a quite detailed picture can be obtained from neutron beam measurements, not by examination of the well-defined diffraction spectra, i.e. the Bragg reflections, which we have been discussing so-far, but by a study of the changes in the general background of the diffraction patterns. Disorder, or departure from the ideal three-dimensional unit of pattern because of impurity atoms, results in neutrons being contributed to this background. By carrying out a Fourier inversion of the magnetic contribution to the background it is possible to determine the spatial distribution of the magnetic disturbance, expressed as a deviation from the moment density in the unperturbed matrix. In fact it proves to be of advantage to abandon the use of neutrons of the usual wavelength of about 1Å in favour of a longer wavelength, say 5Å, which is so long that the interference conditions which lead to the production of the ordinary Bragg reflections can not be satisfied. The background scattering is then predominant and can be more easily examined and analysed in detail. An indication of the type of information which is forthcoming from these experiments can be gained by considering the effect of adding small amounts of impurity to iron or nickel. In the case of iron at least fifteen different metal impurities have been tried, usually in quantities between one and two per cent. In principle, in order that the impurities may function completely as individual atoms, the concentration used should be as low as possible, but this desirable feature has to be balanced against the fact that the concentration of defects has to be high enough to permit accurate measurement of their properties and influence. In these experiments, where it is the background scattering which is being measured, it is not essential to have the high angular resolution which is employed when Bragg reflections are being measured. It is possible therefore to relax the collimation of the incident neutron beam and thus to increase substantially the rather low intensities which the low concentration of impurity atoms would otherwise produce. Measurements with various impurities in iron show that in all cases there is a reduction of the magnetic moment on the impurity site to a value which is lower than that of an iron atom. The extent of the reduction varies widely between different elements. For cobalt the moment falls only from $2.2\mu_B$, the value for iron, to $2.1\mu_B$ but, at the other extreme, titanium and chromium produce an oppositely directed moment, i.e. an antiferromagnetic alignment, of $-0.7\mu_B$. Of possibly more interest is the fact that there are wide differences amongst the various impurity metals concerning the extent to which their disturbing effect spreads outwards to the surrounding matrix of iron atoms. For manganese and titanium the disturbance scarcely extends beyond the impurity site itself but for all the other elements there is a net increase of moment which is distributed amongst the iron atoms which surround the impurity. For V, Cr, Mo, Ru, W, Re and Os, which are elements which lie to the left of or beneath Fe in the Periodic Table, the increased moment is concentrated at a distance of about 5Å from the impurity.This corresponds to about the fourth and fifth neighbours of the impurity and these have their magnetic moments increased by about 0.5%. This may seem a small increase but it must be remembered that there are about 50 neighbours within the spherical shell which extends from 4 to 6Å, so that, overall, there is quite a significant re-distribution of magnetic moment around the impurity atom. The elements Co, Ni, Rh, Pd, Ir and Pt, all of which lie to the right of iron in the Periodic Table, behave rather differently. For them the disturbance of magnetic moment is larger for the nearer neighbours and falls off steadily as we go outwards beyond a distance of 3Å, where it amounts to about 3%. Figure 18 summarises the changes in magnetic moment per unit volume for these two groups of elements. It will be realised that these relatively small changes in moment which have been deduced in this way indicate

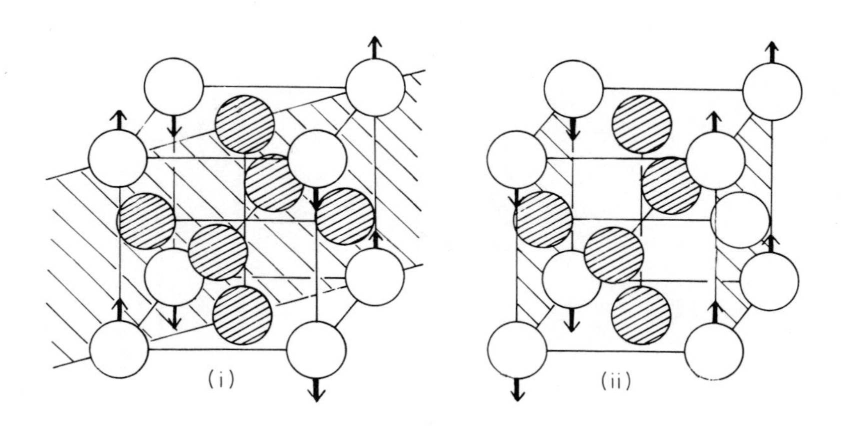

Figure 17 The dependence of the magnetic structure of platinum-iron alloys (in the neighbourhood of the composition Pt_3Fe) on the precise chemical composition. At the stoichiometric composition, Pt_3Fe, and also with less Fe than this indicates, the antiferromagnetic structure indicated at (i) is found, possessing (110) sheets of identically directed moments. When excess of iron is present and there are, necessarily, near-neighbour iron atoms as at A, a new structure appears, shown at (ii), in which the sheets of identical moments are parallel to the cube faces.

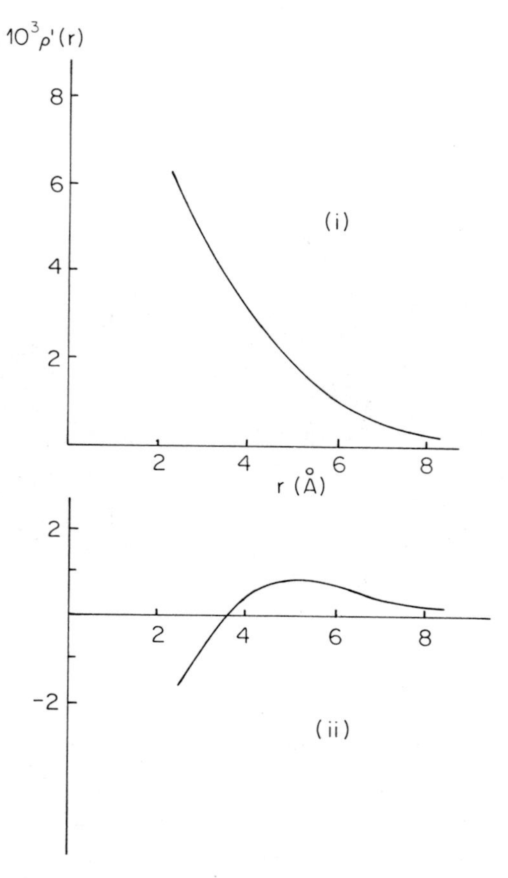

Figure 18 The change in the magnetic moment of the atoms in the iron matrix for dilute alloys in iron, as a function of the distance r (Å) from an impurity atom. If the z axis represents the direction of magnetisation, then ρ'(r) is the deviation of the z component of magnetisation from the value in pure iron. Curve (i) is representative of V, Cr, Mo, Ru, W, Re and Os, which are elements which lie to the left of or beneath iron in the Periodic Table. Curve (ii) applies to Co, N, Rh, Pd, Ir and Pt, which lie to the right of iron in the Table.

the high accuracy which has been attained in the measurement of neutron scattering in these experiments.

Anisotropy of electron distribution: use of polarised neutrons

It is convenient at this point to mention other precise measurements with pure ferromagnetic elements, such as iron, which have the aim of showing whether or not the distribution of magnetic moment in an atom is spherically symmetrical. An alternative view of these experiments is that, referring back to equation 4, they investigate the constancy, or otherwise, of the form-factor curve for different azimuths within the crystal. The conclusions depend on making very precise measurements of the magnitude of the magnetic contribution to the Bragg reflections, for it can be shown that the distribution of magnetic moment in space is the Fourier transform of the scattered amplitude. Moreover these investigations provide a good example of the use of beams of polarised neutrons for making accurate determinations of the magnetic scattering amplitude **p**, particularly for ferromagnetic materials. We digress for a moment to describe the principles of producing and using these polarised beams. In figure 19 let a vertical magnetic field H be applied to a ferromagnetic material which is oriented suitably to give a Bragg reflections from the hkl crystallographic plane, shown shaded in the figure. The vector $\underline{\lambda}$ indicates the direction of polarisation in the incident beam of neutrons. In the neighbourhood of the magnetised material only two orientations of the vector $\underline{\lambda}$ are possible, namely parallel or antiparallel to the field H. It follows from equation 7 that the vector $\underline{q}$ is equal to -1 in magnitude and directed along the field H. Consequently it follows from the dependence on the term $\underline{\lambda}.\underline{q}$ in equation 9 that the scattering amplitude for neutrons whose spin vector $\underline{\lambda}$ is parallel to H will be (b - p) and for those whose vector is antiparallel to H it will be (b + p). Thus the two possible polarisation groups of neutrons will be reflected to different extents. Moreover it is possible to visualise some particular ferromagnetic materials for which the values of b,p are equal. Under these circumstances the value of b - p would be zero and the reflected beam would consist entirely of neutrons for which $\underline{\lambda}$ is antiparallel to H. Thus we would have produced a polarised beam. Several suitable materials are known in practice and a good example is provided by magnetite, for whose 220 reflection b,p are very nearly equal. It will be realised that equality of the two parameters can only hold for a particular reflection because of the dependence of the value of p on the reflection angle θ.

In this way, using a crystal of magnetite as a polariser, a beam of polarised neutrons can be made available for subsequent experiments. The advantage which they provide to the experimenter arises as follows. If the polarised beam, with spin vector directed upwards, is scattered at the plane indicated in figure 20, where a magnetic field has been applied to line up the ferromagnetic moments in an upward direction, then the scattered amplitude will be proportional to (b - p). If the magnetic field is then reversed, so that the moments in the magnetic material are directed downwards, then the scattered amplitude will be (b + p). Thus the ratio of the amplitudes in the two cases will equal (b - p)/(b + p), from which the ratio b/p can easily be deduced. It follows that if, in the case of metallic iron for example, we know the value of the nuclear scattering amplitude b then we shall be able to determine the magnitude of p for any reflection. We have already seen, in equation 4, that the angular variation of p with θ is determined by the form-factor f, so that the above measurements enable us to determine the form-factor curve experimentally and, hence, deduce the distribution around the nucleus of the magnetic electrons present in the atom. The advantage of polarised neutrons is that the measurements can be made much more accurately and can be continued out to relatively large values of θ where the value of f, and therefore of p, is very small. We can understand more easily how this arises if we take a particular case, such as a reflection for which the value of **p** is only one-tenth of b In the polarised-neutron experiments the scattered amplitudes for the two directions of the applied magnetic field will be b - p, b + p respectively and these two quantities will be in the ratio of 0.9/1.1 if p = 0.1b. The experimenter measures intensities, rather than amplitudes, and these will accordingly be in the ratio 0.81/1.21, so that the change of field direction is accompanied by an intensity change of about 50%. On the other hand, when unpolarised neutrons are used for the experiments only very small intensity changes can be expected. Reference to equation 8 will show that no intensity change

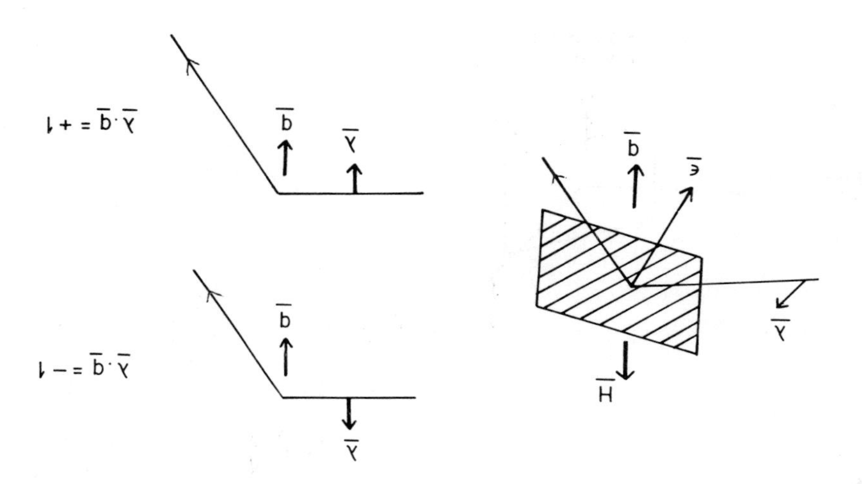

Figure 19 Production of a beam of polarised neutrons by reflection from a suitable ferromagnetic crystal with a vertical magnetic field.

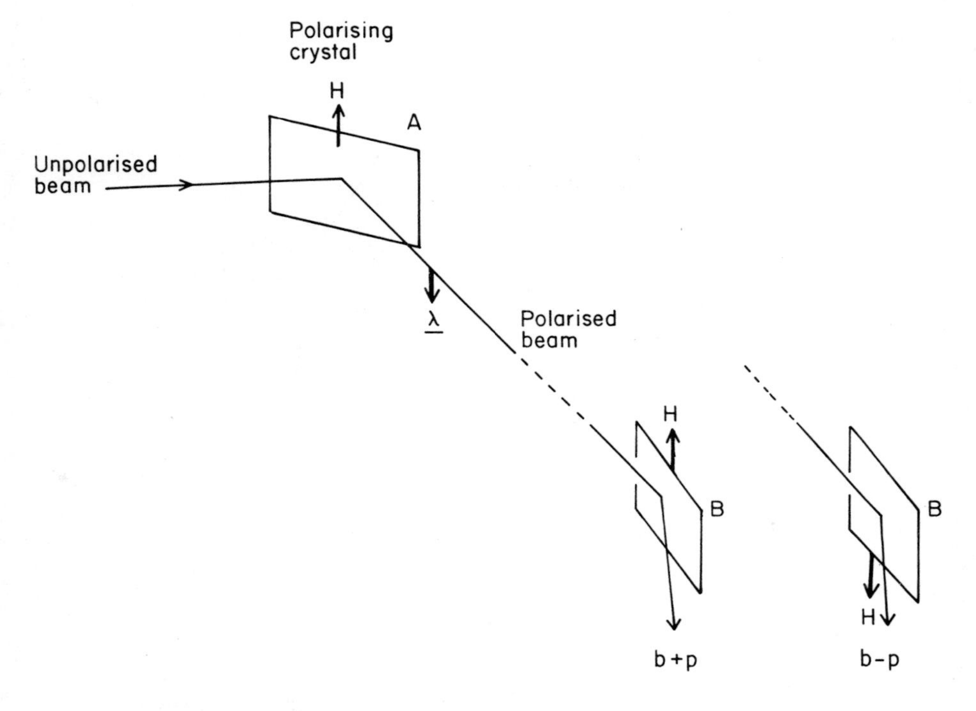

Figure 20 Assessment of small amounts of magnetic scattering from ferromagnetic crystals by measuring the ratio of the scattering observed with upward and downward directed magnetic fields.

occurs when a vertical field is reversed, since the intensity is given by $(b^2 + p^2)$ in each case. The significant observations are the intensities when horizontal and vertical fields are applied in turn and these, from equation 7, will be proportional to b^2 and to $(b^2 + p^2)$ respectively. Insertion of the relation $p = 0.1b$ gives values which are proportional to 1.00 and 1.01 respectively so that the determination of the value of p would rest on the observation and measurement of an intensity change of only 1%. The virtue of the polarised neutrons in this kind of investigation is therefore very evident.

From a practical point of view we must draw attention to an experimental difficulty in the procedure which we have outlined above and indicate a way of overcoming it. We stated that two measurements were to be made in the polarised neutron experiment, corresponding to, first, an upward and, secondly, a downward directed magnetic field in figure 20. In achieving the second of these arrangements there will inevitably be a discontinuity in the magnetic field which will change from upwards at the polarising crystal A to downwards at the sample crystal B. In fact the neutron responds to this discontinuity by reversing its direction of polarisation so that it will arrive at B again with its spin vector aligned with the field. This difficulty can be avoided by following a practice which is well-known in atomic beam work: the field direction at B is always maintained to agree with that at A, but the spin direction of the neutron is turned around by just 180° by applying a suitable axial magnetic field of radio frequency. The arrangement is indicated diagrammatically in figure 21 and by suitable choice of both the magnitude and frequency of this additional field the polarisation of the neutron beam can be turned over with an efficiency of 99%, thus giving the desired accuracy in measuring the ratio of b/p.

As we have described, the experimental measurements for each reflection plane yield a value of p, and hence of the form-factor of the iron atom at the corresponding value of θ. If these values of f are plotted against $(\sin\theta)/\lambda$ they do not lie very satisfactorily on a smooth curve. At high angles in particular it is clear from figure 22 that the departures of the experimental points from the mean curve are much larger than the experimental error. Moreover in certain cases pairs of reflections occur at the same value of θ (these are reflections which have Miller indices hkl which produce the same value of $(h^2 + k^2 + l^2)$, but they are found to display substantially different values of f. This shows that the form-factor is dependent on direction within the crystal and hence that the magnetic electron density is not spherically symmetrical about the nucleus. From the data in the figure 22 it is possible to derive a projection of the spin density on any plane in the crystal and figure 23 illustrates this for the (100) plane. It is evident that the electrons are compressed in the diagonal [110] direction compared with their spread along the cube axes. The anisotropy can be seen more clearly in figure 24 which represents the excess magnetic spin density over that expected from spherical symmetry. From this figure and similar projections on to other planes in the crystal it is evident that there are excesses of electron density along the cube axes and deficiencies along both the face and body diagonals of the cube. Further refinement by Shull and Mook (1966)[10] has revealed interlocking rings of reversed magnetisation centred on points such as ½00 and probably associated with 4s electrons. The integrated contribution of this magnetisation is $-0.21\mu_B$. Metallic nickel has been studied in a similar way (Mook and Shull, 1966 [11] and Mook, 1966[12]) but here there is an excess of electron density along the cube diagonals and a shortage along the cube edges. Between neighbouring nickel atoms there was found a low, almost constant, background of negative magnetisation, integrating to about $-0.1\mu_B$ per atom. On the other hand, for hexagonal cobalt, but not for the face-centred cubic form, the electron density around the atoms is almost spherically symmetrical, but again with an almost constant negative moment density in the regions remote from the atoms. Measurements have also been made with many of the rare-earth metals, but the study of some of these is made difficult by the enormous coefficients of absorption which they have for neutrons. Thus samarium and gadolinium, which in the natural state have absorption cross-sections of 3500 barns (10^{-24}cm^2) and 20,000 barns respectively, have proved amenable to study in the form of single-crystals of ^{154}Sm and ^{160}Gd. Samarium is interesting because the S,L vectors are oppositely directed and this leads, in the nomenclature of equation (6),

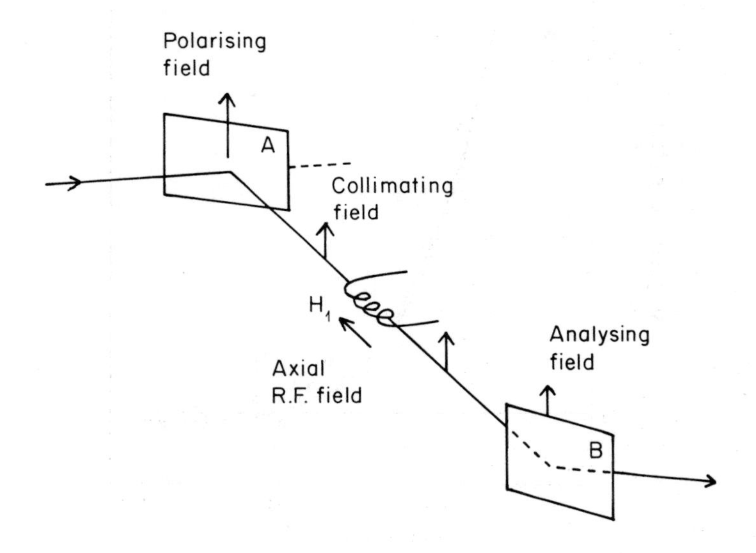

Figure 21 A satisfactory experimental method for reversing the spin-moment/field-direction relationship by applying a suitable radio-frequency field.

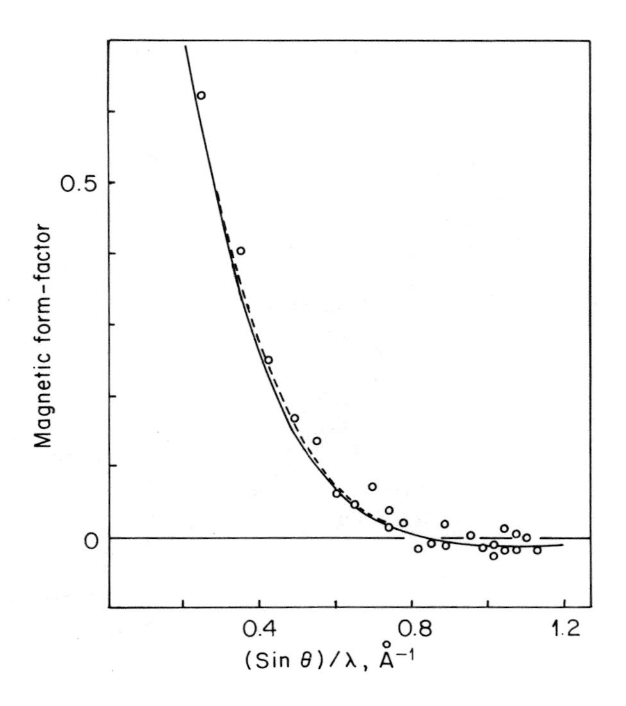

Figure 22 The variation of the form-factor of iron with (sinθ)/λ indicating its dependence on the orientation in space of the reflection plane, caused by the anisotropic distribution of the magnetic electrons in the iron atoms (Shull and Yamada, 1962)[9]. The circles indicate experimental points and the curves are the results of two alternative calculations.

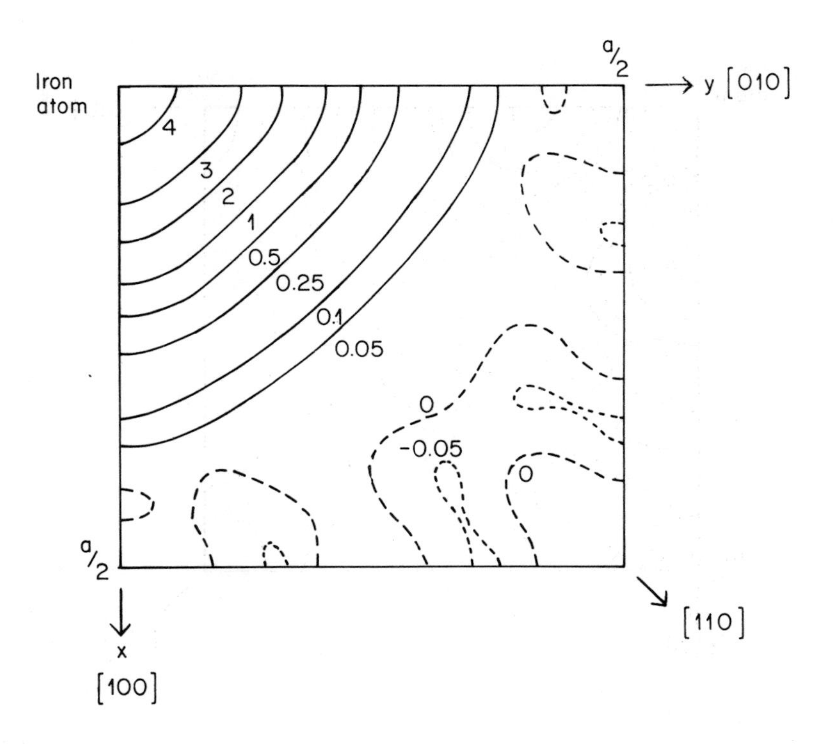

Figure 23 A plot of the distribution of magnetic moment in the basal plane of the unit cell of iron, indicating the compression of the electrons along the diagonal [110] direction compared with the direction of the cube axis [100] (Shull and Yamada, 1962)[9].

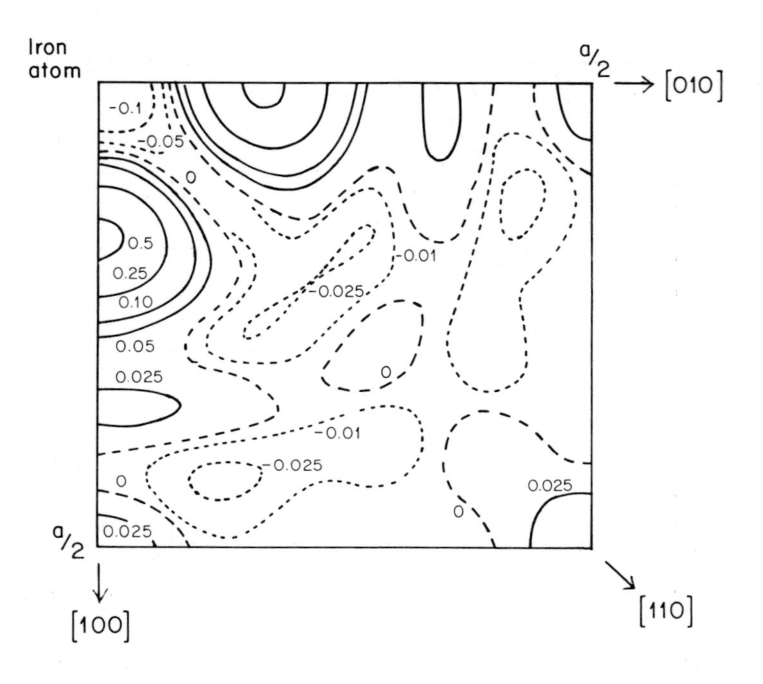

Figure 24 A more direct indication of the anisotropy of the electron distribution in iron, obtained by subtracting from the experimentally-deduced value of the electron density the value which would be expected if spherical symmetry did occur (Shull and Yamada, 1962)[9]

to coefficients of f_S, f_L which are in the ratio of -5 to +6 and results in a form factor which peaks at a finite value of $(\sin\theta)/\lambda$ equal to about 0.4 $Å^{-1}$. This is justified by figure 25 which shows the results of Koehler and Moon (1972)[13] who have determined the antiferromagnetic structure of samarium. Under these circumstances, where the form-factor curve is falling steeply towards $\theta = 0^o$ it is not possible to determine the value of the magnetic moment of samarium very accurately but it is clearly small and probably close to $0.1\mu_B$. It is suggested that the free-ion moment of $0.71\mu_B$ is much reduced by a large polarisation of the conduction electrons parallel to the ionic spin. Gadolinium is a hexagonal close-packed ferromagnetic for which L = O and S = J. Measurements with both polarised and unpolarised neutrons by Moon et al. (1972)[3] have shown that the density of the 4f electrons is spherically symmetrical but there is a diffuse component of unpaired spin due to the conduction electrons which has the unusual distribution shown in figure 26. The density is negative within the undulating columns parallel to the c-axis which run through the C sites of the hexagonal close-packed structure.

Covalency

The foregoing discussion of the determination of magnetic form-factors has mainly been concerned with ferromagnetic metals, for which very accurate measurements can be made with polarised neutrons. Metallic salts and oxides, such as MnO, MnS, and MnF_2, are generally antiferromagnetic and were among the first magnetic materials to be examined by neutron diffraction. Later, more accurate, measurements on such materials - particularly attempts to determine the form-factors of the magnetic ions - have indicated the application of neutron diffraction to the study of covalency. If the bond in NiO, for example, was completely ionic then there would be complete transfer of two electrons from nickel to oxygen. In fact the transfer is not quite complete and we may consider that a fraction of an electron, with spin oppositely-directed to that on the Ni^{2+} ion, is transferred back from oxygen to nickel. The outcome is that there is a small amount of unpaired spin moment on the oxygen ligand and, also, the magnetic moment on the nickel ion is slightly reduced below its expected value to $1.81\mu_B$ (allowing for some orbital contribution the moment would be expected to be $2.23\mu_B$). In an actual magnetic material each cation is surrounded by several anions, often octahedrally, and each of these will indulge in charge transfer and contribute to the reduction of the spin moment on the cation. In many antiferromagnetic materials (such as MnO, NiO) an anion may have cation neighbours whose moments are oppositely directed and the anion will not then show any net spin: in other cases however, such as MnF_2 where the anion environment is not symmetrical, a net spin will be built up on the anion. In either case the details of the resulting spin distribution will influence the magnetic scattering of the material and, particularly, the effective form factor of the magnetic ion. The situation may be summarised by figure 27. Transfer of the oppositely-directed spin results in a reduction of moment, indicated by the fall of the point A below unity and a very slight broadening of the form-factor. For a paramagnetic ion or in a ferromagnetic material the net charge on the ligand adds a forward-going peak to the curve which restores it to unity at $Q = 0$, but this peak is usually below the θ value of the first coherent Bragg reflection and therefore not evident: in most antiferromagnetic materials the ligand peak does not occur. In the more usual antiferromagnetic structure, such as MnF_2, the existence of the ligand charge leads to anomalous magnetic intensity in reflections which would otherwise be expected to be purely nuclear. Such reflections can be assessed accurately with polarised neutrons. For a recent review of the application of magnetic scattering to the study of covalency and a comparison with results from resonance techniques the reader is referred to an article by Tofield (1975)[16].

Atomic Dynamics

Our discussions so far have been limited to what we may call a 'static picture' of the magnetic structure of materials. This is only one aspect of a more complete picture of magnetism. The atoms in a solid, whether it has magnetic properties or not, are not stationary. They possess kinetic energy and are in motion about their equilibrium positions. These motions are not independent, because the motion of any atom will affect its distances from its neighbours and hence the interatomic forces. In fact the instantaneous position of an individual atom is the result of the motions arising from a whole

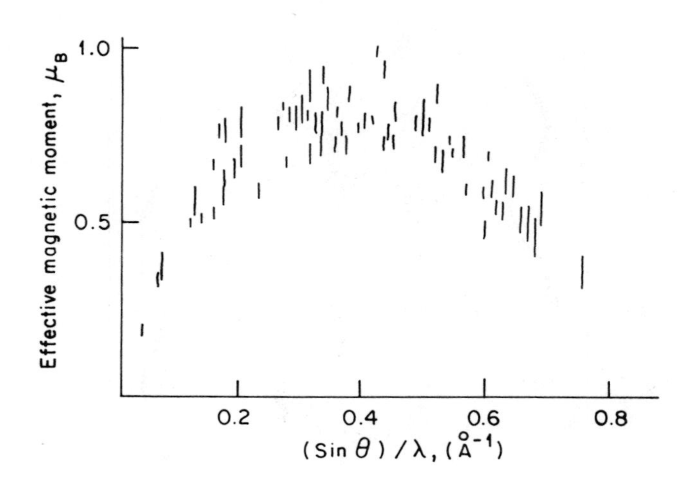

Figure 25 The form-factor for metallic samarium deduced experimentally, indicating that maximum scattering occurs when $(\sin\theta)/\lambda$ is equal to about 0.4 $\mathrm{\AA}^{-1}$. The ordinate scale indicates the effective magnetic moment in μ_B (after Koehler and Moon, 1972)[13].

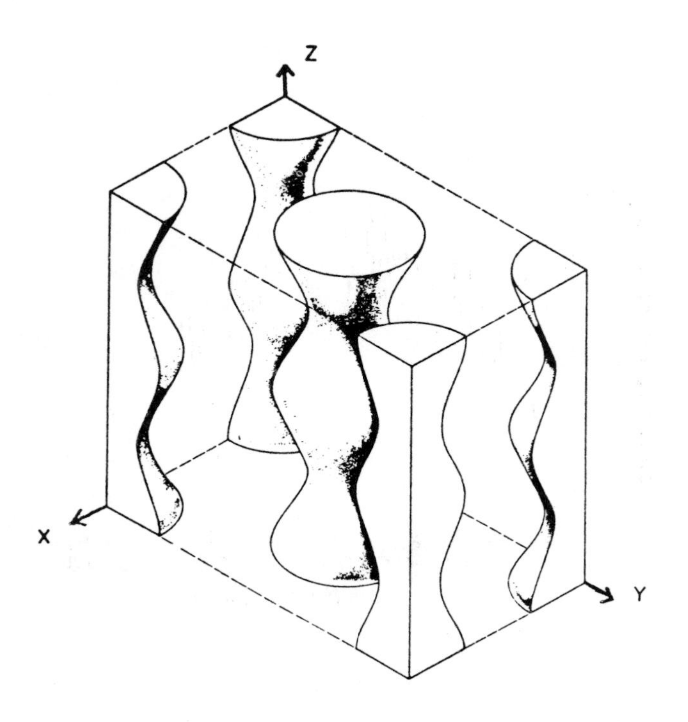

Figure 26 The distribution of the diffuse component of unpaired spin-density in the unit cell of gadolinium. Within the pillar-like figures the density is negative Moon et al. 1972)[14].

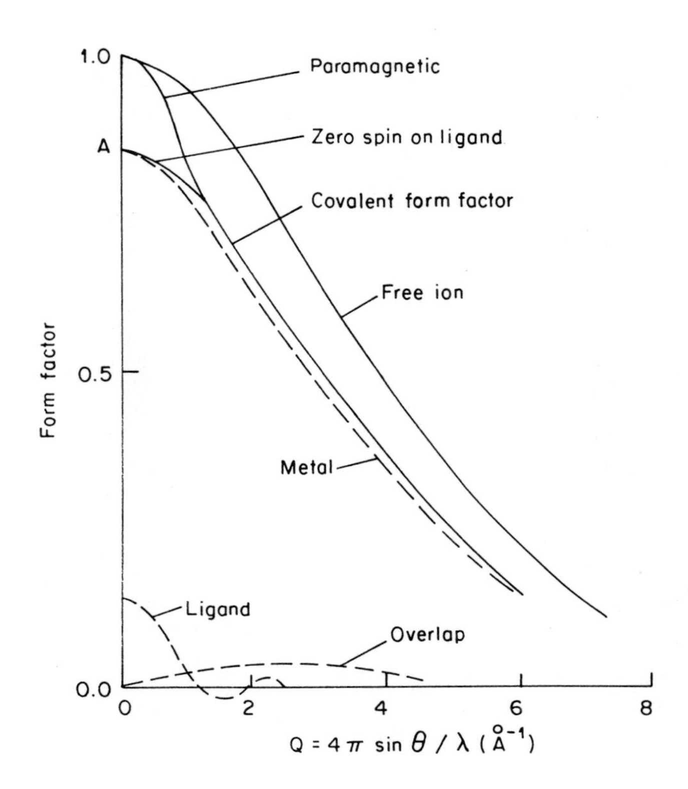

Figure 27 The effective of covalency on the form-factor of a magnetic cation. For most antiferromagnetic materials the curve 'terminates' at A, giving a reduced moment. In MnF_2 there is a net charge on the ligand and the curve is restored to unity with a sharpened peak at Q = O (Jacobson, 1973)[15]

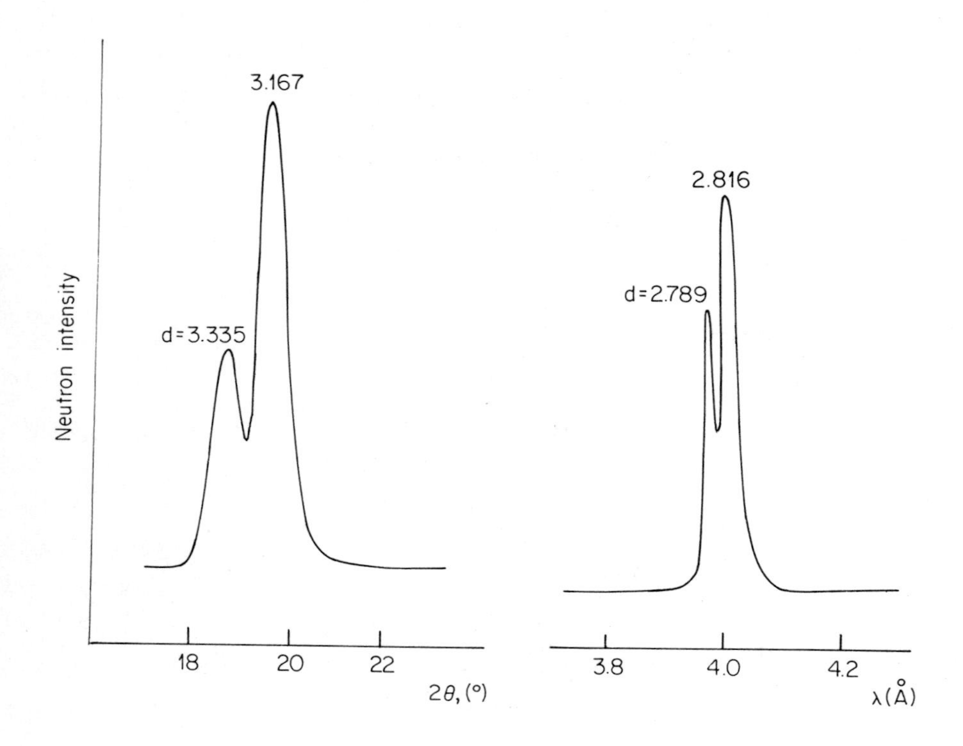

Figure 28 A comparison of the resolution which can be obtained for reflections of closely similar interplanar spacing, for conventional and time-of-flight methods of diffractometry. For the former, a change of spacing from 3.335 to 3.167 Å (roughly 5%), gives a pattern of similar resolution to that which the latter achieves for a spacing change from 2.789 to 2.816 Å, which is only 1%.

spectrum of vibrations which are being propagated through the crystal. In a magnetic material the magnetic structure is also dynamic: the magnetic spins move with the atoms and they are also in angular oscillation to some extent. If a magnetic moment is displaced or rotated then a wave of magnetic disturbance will propagate this change through the crystal. In the same way as, in quantum terms, we speak of phonons in an ordinary crystal, we speak of magnons in a magnetic material. The phonon spectrum of solids has been studied by neutron methods for many years: study of the "spin-wave" spectrum has made rather less progress, mainly because the magnetic scattering is generally weaker than the nuclear scattering and has to be separated out from the latter before it can be examined at all. In each case the study depends on the assessment of inelastic scattering, in which a substantial interchange of energy has taken place between the incident neutrons and the phonons or magnons in the solid. Determination of the energy change is achieved by measuring the wavelength distribution of the scattered neutrons and this procedure results in low intensities, so that rather large crystals are needed for study, even with high-flux reactors. Consequently the rate of progress in this kind of work is slow. Nevertheless, the potentialities of this research are very great in relation to the search for a thorough fundamental understanding of magnetism. Thus very distinctive differences have been demonstrated between the dispersion laws for magnons in ferromagnetic and antiferromagnetic substances. For the former, for example in $Co_{0.92}Fe_{0.08}$ which was chosen for study by Sinclair and Brockhouse (1960)[17] because of the small nuclear scattering of the cobalt atoms, the frequency ω of the spin waves is proportional to q^2, i.e. the square of their wave-number, for small values of q[†]. For antiferromagnetic materials, on the other hand, there is a linear relation between ω, q which was convincingly demonstrated by Windsor and Stevenson (1966)[18] for $RbMnF_3$. Much interest has also concentrated on the fluctuations of magnetic order which occur in the neighbourhood of the Curie or Néel temperature, giving rise to the so-called critical scattering. For

† q here is the wave-number of the spin waves and must be distinguished from the magnetic interaction vector of equation (7).

a recent study and further references on this topic the reader is referred to a paper by Passell, Als-Nielsen and Dietrich (1972)[19] concerning europium oxide, EuO, chosen as a good representation of a simple, isotropic Heisenberg ferromagnet.

Some alternative techniques

We mentioned at an early stage in this article that the principle of producing a monochromatic beam of neutrons, from the heterogeneous velocity spectrum emerging from the reactor, was very uneconomical, using, as it did, only about 1% of the vailable neutrons. In an alternative technique, which has been investigated by several researchers in recent years, the whole of the spectrum falls on the sample under investigation and the diffracted neutrons are examined only at a single value of θ. The neutrons which arrive at the observation point are sorted in wavelength by a time-of-flight technique, so that in the Bragg equation $\lambda = 2d \sin \theta$ the dependent variables are λ, d instead of θ,d as is the case in the conventional monochromatic technique. A further significant advantage of the time-of-flight method is that it is much easier to get good angular resolution at large values of interplanar spacing. This is illustrated in figure 28 . The left hand pattern for an alloy AuMn indicates that there is reasonable resolution between two diffraction peaks corresponding to spacings of 3.335 and 3.167 Å in a conventional monochromatic beam pattern. This difference of spacing amounts to approximately 5%. The right hand pattern in the figure shows how equally good resolution is obtained in a time-of-flight pattern for $BiFeO_3$ for peaks of 2.789, 2.816 Å which is only a 1% change of spacing. On the other hand, quantitative interpretation of the relative intensities of the reflections from different planes is much more difficult in the time-of-flight technique, simply because the reflections originate from neutrons of different wavelength and it is necessary to know accurately the wavelength spectrum of the incident beam and also the precise dependence of crystal-reflectivity upon wavelength. However, two other advantages of this technique are noteworthy. Time-of-flight methods are particularly applicable to pulsed beams, either from a pulsed reactor or from linear electron accelerators in which the electron beam falls on a target of a heavy element, and this offers immense possibilities for the study of transient effects and relaxation phenomena. At the same time the fact that a fixed angular position is chosen for the counter means that magnets, cryostats and furnaces for controlling the temperature and environment of the sample can remain stationary, thus promising considerable simplification of construction. It is anticipated that the time-of-flight technique will be developed further in the future, bearing in mind the possibility that future increases in neutron intensity may depend on the use of pulsed sources.

As another technique which has been much developed in recent years we mention the study of small-angle scattering, which has been much encouraged by the appearance of special instruments such as that at Grenoble, attached to neutron guide tubes and capable of observations at scattering angles of only a few minutes of arc. Such instruments employ wavelengths up to about 20 Å, using neutrons derived from a cold-source within the reactor, and can examine structural effects of magnitude in the range from 50 - 5000 Å. An application of magnetic interest is the study of the ferromagnetic precipitates which grow in single-crystals of copper containing 1% of cobalt by Ernst et al. (1971)[20]. In such a case the contour of the scattered beam is given by

$$I = I_o \exp(-Q^2R^2/3) \qquad (14)$$

where $Q = 4\pi\sin\theta/\lambda$ and **R** is the effective radius of the diffracting particles. In the work referred to it is concluded that the precipitates have a volume corresponding to spheres of about 150 Å radius.

Finally, we emphasise that we have considered almost exclusively those magnetic materials which have a co-operative arrangement of magnetic spins and we have done this because neutron beams provide such a powerful method of investigating these structures. However, it is to be noted that some work has been done with paramagnetic materials. As we have already pointed out, in paramagnetic substances there is no correlation between the directions of the magnetic moments on neighbouring atoms and the result of this is that the magnetic scattering makes no contribution to the general background of the diffraction pattern and it can be shown that the scattering cross-section per unit solid angle $\frac{d\sigma}{d\Omega}$ is given by the relation

$$\frac{d\sigma}{d\Omega} = \frac{2}{3} S(S+1) \mu_o^2 \left(\frac{e^2\gamma}{m}\right)^2 f^2 \qquad \left[\frac{d\sigma}{d\Omega} = \frac{2}{3} S(S+1)\left(\frac{e^2\gamma}{mc^2}\right) f^2\right]_{emu} \tag{15}$$

where f is the form factor which we have discussed previously. It will be seen that the magnitude of $\frac{d\sigma}{d\Omega}$ is roughly equal to p^2, the distinction being that the factors S^2 in the expression for p^2 has been replaced by $\frac{2}{3}$ S(S+1) and J^2 by $\frac{2}{3}$ J(J+1) for the rare-earth elements. This change takes account of the fact that there is no defined direction for the magnetic spins in a ferromagnetic material and it is possible for the spin direction to be altered as a result of the encounter with the neutron. In some cases it is possible to calculate fairly accurately the contribution to the background scattering from other causes, such as thermal diffuse scattering and nuclear spin or isotope incoherence, and it is then possible to deduce the value of S(S + 1) by using equation 15. Atoji (1961, 1970)[21,22] has made many measurements of this kind among the carbides of the rare-earth elements, using his measured values of J to interpret the valency states of the ions in these compounds.

References

1. H.M. Rietveld, J. Appl. Crystallogr., 2, 65-71 (1969)
2. M. Blume, A.J. Freeman and R.E. Watson, J. Chem. Phys., 37, 1245-53 (1962)
3. R.M. Moon, T. Riste and W.C. Koehler, Phys. Rev., 181, 920-31 (1969)
4. W.C. Koehler in "Magnetic Properties of Rare-Earth Metals" ed. J. Elliott pp. 81-128 Plenum Press, London
5. W.C. Koehler, E.O. Wollan and M.K. Wilkinson, Phys. Rev., 118, 58-70 (1960)
6. G.E. Bacon and J. Crangle, Proc. Roy. Soc., A272, 387-405 (1963)
7. G.E. Bacon and E.W. Mason, Proc. Phys. Soc., 92, 713-25 (1967)
8. M.F. Collins and G.G. Low, Proc. Phys. Soc., 86, 535-48 (1965)
9. C.G. Shull and Y. Yamada, J. Phys. Soc., Japan 17, Suppl. B-III, 1-6 (1962)
10. C.G. Shull and H.A. Mook, Phys. Rev. Lett., 16, 184-6 (1966)
11. H.A. Mook and C.G. Shull, J. Appl. Phys., 37, 1034-5 (1966)
12. H.A. Mook, Phys. Rev., 148, 495-501 (1966)
13. W.C. Koehler and R.M. Moon, Phys. Rev. Letters, 29, 1468-72 (1972)
14. R.M. Moon, W.C. Koehler, J.W. Cable and H.R. Child, Phys. Rev., B5 997-1016 (1972)
15. A.J. Jacobson in "Chemical Applications of Thermal Neutron Scattering" ed. B.T.M. Willis, Oxford University Press, Oxford (1973)
16. B.C. Tofield, Structure and Bondong, 21, 1-87 (1975)
17. R.N. Sinclair and B.N. Brockhouse , Phys. Rev., 120, 1638-40(1960)
18. C.G. Windsor and R.W.H. Stevenson, Proc. Phys. Soc., London 87, 501-4 (1966)
19. L. Passell, J. Als-Nielsen and O.W. Dietrich, "Neutron Inelastic Scattering" pg. 619-629, I.A.E.A. Vienna (1972)
20. M. Ernst, J. Schelten and W. Schmatz, Phys. Stat. Solidi (a) 7, 469-76 (1971)
21. M. Atoji, J. Chem. Phys., 35, 1950-60 (1961)
22. M. Atoji, J. Chem. Phys., 52, 6430-3 (1970)

BIBLIOGRAPHY

Bacon G.E. "Neutron Diffraction" 3rd Edition. Clarendon Press, Oxford (1975), a general account.

Egelstaff P.A., "Thermal Neutron Scattering", Academic Press, London (1965), a specialist account for research workers.

Gurevich I.I. and Tarasov L.V. "Low Energy Neutron Physics", North Holland, Amsterdam (1968), concentrating mainly on the nuclear scattering of slow neutrons.

Izyumov Yu.A. and Ozerov R.P. "Magnetic Neutron Diffraction", Plenum Press, New York (1970).

Marshall W. and Lovesey S.W. "Theory of Thermal Neutron Scatting", Clarendon Press, Oxford (1971).

Neutron Investigation of Magnetic Densities and Magnetic Excitations in Solids

A. PAOLETTI
University of Rome

Introduction

Thermal neutron scattering provides an essential tool for the study of static and dynamic properties of magnetic materials at microscopic level. Thermal neutrons act as very sensitive magnetic probes because of their basic properties which turn out to be extremely valuable in scattering experiments:

1. wavelength comparable to interatomic distances in crystals, so that interference effects arise,
2. magnetic moment which leads to appreciable interaction with magnetic systems,
3. energy of the same order of magnitude as the energy characteristic of magnetic excitations

Magnetic properties of solids are determined by unpaired electrons. Their static distribution at the sites of the unit cell i.e. the "magnetic structure" is closely related to typical macroscopic behaviour which in first approximation might be considered paramagnetic, ferromagnetic or antiferromagnetic. However for a better understanding of the origin of the magnetic properties and a meaningful comparison with theory, information about the unpaired electron distribution in the whole volume of the unit cell, the "magnetization density", as it can be gained through magnetic elastic scattering of neutrons, is required. The picture of a magnetic system must be completed by the analysis of magnetic excitations which provides information on the dynamical behaviour of unpaired electrons. By neutron inelastic scattering the spectrum of magnetic excitations (spin waves) in a significant energy and momentum range can be determined.

It must be recalled that in both the elastic and inelastic scattering experiments, magnetic scattering is associated with nuclear scattering from the nuclei present in the crystal structure and from the crystal thermal excitations (phonons). It is also to be expected that in magnetization densities and spin wave investigations, instrumental resolution plays a major role. In a neutron scattering experiment, the right interpretation of results is possible only if the experimental conditions are clearly understood and their implications correctly evaluated.

While for the basic principles of neutron scattering and

instrumentation, reference is made to the chapter "Neutron Diffraction" by G.E. Bacon, and for a deeper theoretical insight to the book by Marshall and Lovesey[B4], the present chapter deals mostly with typical experimental methods in neutron scattering studies of magnetization density and spin waves, in the temperature range away from magnetic transitions. Critical scattering contribution can then be disregarded. A brief review of significant results is also included providing typical examples of the information which can be gained through these experiments and giving possible suggestions for the correct approach to be taken for solving future problems.

The magnetization density in a magnetic system

The magnetization density in magnetic materials is due to unpaired electrons mostly belonging to outer shells and is simply related to the form factor of magnetic neutron scattering.

For elastic neutron scattering from an atom, the coherent cross section can be written, under the assumption of quenching of orbital moment

$$\frac{d\sigma}{d\Omega} = b^2 + 2bp\underline{q}.\underline{\lambda} + q^2p^2 \quad (1a)$$

$\frac{d\sigma}{d\Omega}$ = differential neutron cross section

b = coherent nuclear scattering amplitude

$\underline{\lambda}$ = unit vector in the polarization direction of neutrons

$\underline{q}$ = $\underline{k}(\underline{k} \cdot \underline{m}) - \underline{m}$ is the magnetic interaction vector

$\underline{k}$ = scattering vector

$\underline{m}$ = unit vector in the direction of the atomic moment

p is the coherent magnetic scattering amplitude and is given by

$$p = \mu_o \frac{\gamma e^2}{2m} gSf(\underline{k}) \qquad \left[p = \gamma \frac{e^2}{2mc^2} gSf(\underline{k})\right]_{emu} \quad (1b)$$

where γ is the magnitude of the neutron moment in nuclear magnetons (= 1.913),

e and m the electron charge and mass respectively,

gS is the atomic moment in Bohr magnetons and

$f(\underline{k})$ is the magnetic form factor for an atom normalized to unity for forward scattering ($\underline{k}$ = 0) and is defined (neglecting the orbital contribution to the magnetic moment) as the Fourier transform of the normalized spin density $s(\underline{r})$ (which for most calculations can be treated as a scalar quantity i.e.

$$f(\underline{k}) = \int \exp(i\, \underline{k} \cdot \underline{r})\, s(\underline{r})\, d\underline{r} \quad (2)$$

where $\underline{k} = \underline{K} - \underline{K}'$ is the scattering vector i.e. the difference between the incident and the scattered wave-vector , and the integral is extended to all space).

This relationship is equivalent to the one for the X-ray scattering factor, $s(\underline{r})$ being replaced by the charge density $\rho(\underline{r})$. In scattering experiments by solids, the intensity at Bragg angles is determined by the structure factor F(hkl) which represents the sum of contributions to the coherent scattering by the various atoms of the unit cell. It can be also shown that charge or magnetization densities can be determined at any point of the unit cell by measuring the elastic X-ray scattering or neutron magnetic scattering respectively, at the Bragg positions only.[B3] Accordingly the density is given by the relationship

$$\sigma(x,y,z) = \frac{1}{V} \sum_{hk\ell}{}_{-\infty}^{+\infty} F(hkl) \exp\left[-2\pi i\left(\frac{hx}{a} + \frac{ky}{b} + \frac{lz}{c}\right)\right] \tag{3}$$

where $\sigma(x,y,z)$ is the density and F(hkl) is the structure factor for the Bragg reflection (hkl). As the sum is over all the (hkl)'s the structure factor for all the reflections would be measured. This is impossible of course and with X-rays only charge densities in limited cases would be obtained. On the other hand in magnetic neutron scattering, use can be made of (2) as well: $\sigma(x,y,z)$ in this case is the magnetization density and F(hkl) is the magnetic structure factor. The magnetization density is assumed to be everywhere collinear so that σ is a scalar function. If orbital contribution to the magnetic moment is present, $f(\underline{k})$ becomes a combination of spin and orbital form factors.

For a single atom the neutron magnetic form factor can be also expressed, in absence of orbital contribution, by

$$f(\underline{k}) = \int [A\,\rho_{\uparrow}(\underline{r}) - B\,\rho_{\downarrow}(\underline{r})] \exp(i\underline{k} \cdot \underline{r})\,d\underline{r} \tag{4}$$

where $\rho_{\uparrow}(\underline{r})$ and $\rho_{\downarrow}(\underline{r})$ are the charge density for electrons of spin up and down respectively. Therefore magnetic form factors play a relevant role in the determination of charge density in solids. If the charge density is spherically symmetric, then (4) reduces to

$$f(k) = \int r^2 [R^2_{\uparrow}(r) - R^2_{\downarrow}(r)] \frac{\sin kr}{kr}\, dr \tag{4'}$$

where R^2 is the radial part of the charge density. Equation (4') says that for spherically symmetric charge distribution the form factor is independent of the direction of $\underline{r}$, so that any pair of reflections of the same $|\underline{r}|$ must exhibit the identical form factor.

It must be pointed out that magnetic neutron scattering is a very powerful direct tool for determining the distribution of unpaired electrons whose contribution to X-ray scattering is in general heavily masked by the contribution of the closed electronic shells. On the contrary the paired electrons of the closed shells are not "seen" by neutrons as the volume occupied by them has zero magnetization. Unpaired electron distribution can then be determined with appreciable accuracy. It has been possible, in this way, to gain valuable experimental information on electronic wave-functions for 3d, 5d, 4f electrons, to be compared with theoretical values.

Form factor measurements provide therefore through equation (3) the magnetization density $\sigma(x,y,z)$ which is caused by unpaired electrons. In calculating $\sigma(\underline{r})$ from experimental neutron data the termination problem arises as the decrease of intensity of magnetic scattering with $\underline{k}$ does not permit to include many terms in the sum over hkl appearing in equation (3). In Figure 1 a simple approximation to the 3d radial charge density in Fe, $\rho(r)$ is given. Now, if scattering factors are calculated according to this charge density, it is possible to eliminate from equation (3) the amount of the termination error. The results are given in Figure 2 which provides the $\rho(r)$ calculated at 1/8, 1/8, 1/8 (half the nearest-neighbour midpoint) for an increasing number of terms in the Fourier sum. One sees that only including all the terms up to $\frac{\sin\theta}{\lambda} = 2.5\ \text{Å}^{-1}$ a reasonable accuracy of, say, 5% is obtained. Now it is well known that magnetic scattering at such values of $\frac{\sin\theta}{\lambda}$ is exceedingly small and an accurate experimental determination of all the magnetic structure factors up to $\frac{\sin\theta}{\lambda} = 2.5$ is then impossible. However meaningful information can be obtained of also for values of $\frac{\sin\theta}{\lambda}$ up to $1\ \text{Å}^{-1}$ only, provided one looks for the average values in a small cube of edge length 2δ and volume $V = 8\delta^3$ centred at x,y,z [2] instead of requiring the charge density or the magnetization density at a specific point x,y,z. One obtains

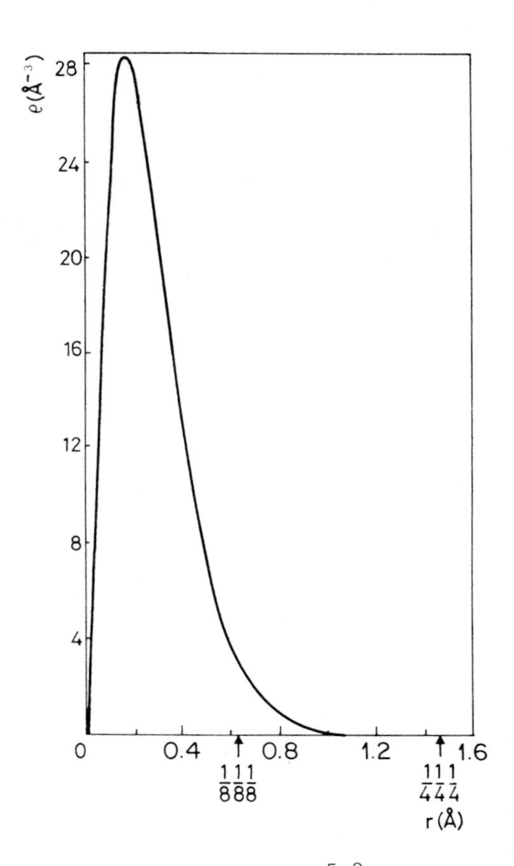

Figure 1 The radial charge density = $(a^5 r^2/24) \exp(-\alpha r)$ as a function of r for a = 10.8 Å. This charge density is a simple approximation to the 3d radial charge density in Fe (from R.J. Weiss: (B6) p. 61).
(Reproduced by permission of North-Holland Publishing Company)

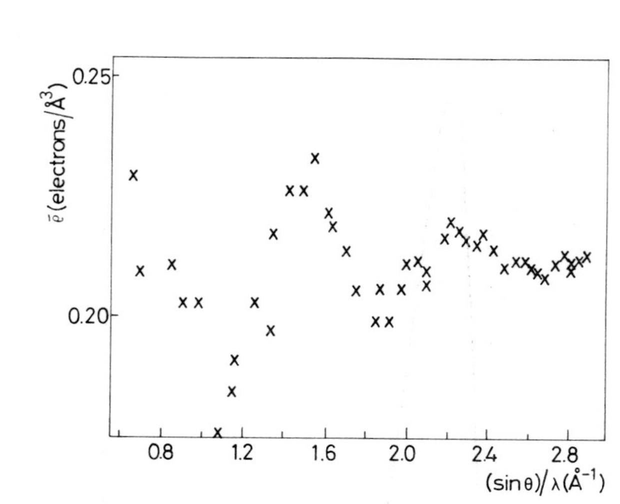

Figure 2 The charge density at the 1/8, 1/8, 1/8 position in the hypothetical b.c.c. unit cell (lattice parameter 3 Å and scattering factor based on the radial density given in Figure 1) as a function of increasing the number of reflections in the Fourier sum (from R.J. Weiss (B6) p. 76).
(Reproduced by permission of North-Holland Publishing Company)

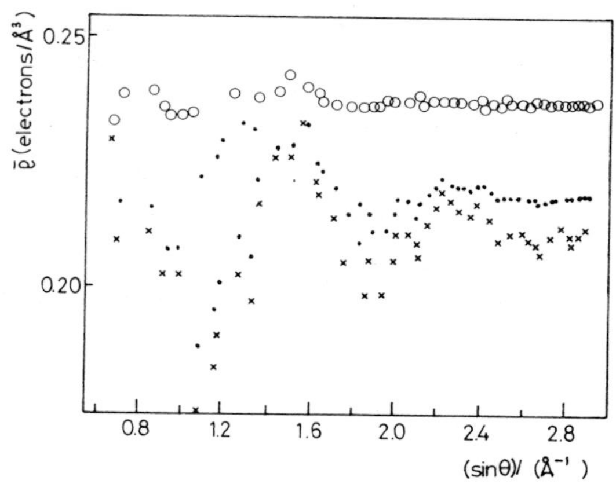

Figure 3 Identical to Figure 2 (crosses) and charge density averaged over a cube centred at 1/8, 1/8, 1/8 and cube sizes δ/a = 0.005 and 0.10 (points and circles respectively) (from R.J. Weiss (B6) p. 76).
(Reproduced by permission of North-Holland Publishing Company)

$$\underline{\sigma}(x,y,z) = \int \frac{1}{8\delta^3} \int_{x-\delta}^{x+\delta} dx \int_{y-\delta}^{y+\delta} dy \int_{z-\delta}^{z+\delta} dz \quad \sigma(x,y,z) \tag{2'}$$

$$= \frac{1}{V}\sum_h \sum_k \sum_l \frac{\sin 2\pi h\delta}{2\pi h\delta} \; \frac{\sin 2\pi k\delta}{2\pi k\delta} \; \frac{\sin 2\pi l\delta}{2\pi l\delta} \; F_{hkl}[\exp 2\pi i(hx+ky+lz)]$$

The convergence of equation (2') for increasing values of (hkl) is much faster than the convergence of series (3) because of the presence of the factor $(hkl)^{-1}$, as can be seen in Figure 3.

On a similar principle is based a method for calculating the magnetic moment associated to a crystallographic site without requiring bulk magnetization measurement and form factor extrapolation to $\frac{\sin \theta}{\lambda} = 0$.

Important information on electron distribution can be also gained from experimental determination of the form factor at positions different from Bragg peaks. If it is possible for instance, by measuring the form factor at very small angular intervals, to obtain an appreciable portion of the curve in a practically continuous fashion; the overall form factor curve can then be deduced from (3) by best fitting procedure.

It is convenient of course to limit the "continuous measurements" to that $\underline{k}$ interval corresponding to the $\underline{r}$ interval where a particular accuracy in the determination of magnetization density is required. For other $\underline{r}$ the information obtained from Bragg peaks will be used.

Furthermore it has been pointed out [Bl] that possible differences between form factors for the itinerant electron model and the localized model, can be observed away from Bragg positions. In other words, at a general position in reciprocal space the form factor is not necessarily a smooth interpolation of the values measured at reciprocal lattice points. Also if the complete equivalence of the two form factors would be ascertained, form factors measured at $\underline{k}$ smaller than the first Bragg peak would be extremely valuable as they provide information on magnetization density at high $\underline{r}$ where the contribution of conduction electrons is dominant.

Experimental methods

From equation (3) one sees that the evaluation of magnetization densities requires the measurement of magnetic structure factors F^m_{hkl} for all reflections up to a certain $\frac{\sin\theta}{\lambda}$. In order to obtain the best accuracy in the experimental data, one looks, in general for the method which provides the highest sensitivity for the system being investigated. It is then important to consider advantages and disadvantages of the different methods keeping in mind that high sensitivity is sometimes associated to low precision.

The experimental data on magnetization densities are taken normally by using a double axis neutron spectrometer for determining the magnetic structure factors from Bragg peaks. According to the problem, polarized or unpolarized neutron beams are used. Information on form factors at $\underline{k}$ values different from Bragg positions, can be obtained by measuring the incoherent elastic magnetic scattering or by measuring the magnetovibrational inelastic scattering with a triple axis spectrometer as will be discussed later.

Magnetization densities by elastic scattering of polarized neutrons

A double axis spectrometer modified for polarized neutrons is used. As discussed in the chapter "Neutron Diffraction" the use of polarized neutrons enhances the magnetic contribution to the scattering and higher sensitivity in determining the magnetic structure factors can be obtained for ferromagnets and some anti-ferromagnets.

For polarized neutrons and experimental geometry with $\underline{\lambda}$ parallel to $\underline{m}$, both vectors being perpendicular to $\underline{k}$.

$$\frac{d\sigma}{d\Omega} = (b \pm p)^2 \qquad (3)$$

From Figure 4 the advantage of using polarized neutrons is evident. In order to determine the magnetization density it is necessary to measure with the highest possible accuracy the magnetic structure F^m_{hkl} . This can be done by determining the so called "polarization ratio" R, that is the ratio, for each Bragg peak, of the scattered intensity for incident neutrons polarized respectively parallel and antiparallel to the magnetization of the sample. The spectrometer must be provided with a polarizer and a spin flipper device (Figure 5). One gets

$$R_{hkl} = \frac{P(F^n_{hkl} + F^m_{hkl})^2 + (1-P)(F^{n^2}_{hkl} + F^{m^2}_{hkl})}{P\,\phi(F^n_{hkl} - F^m_{hkl})^2 + P(1-\phi)(F^n_{hkl} + F^m_{hkl})^2 + (1-P)(F^{n^2}_{hkl} + F^{m^2}_{hkl})}$$

$$= \left(1 + \frac{2P\gamma}{1+\gamma^2}\right) \Big/ \left(1 - \frac{2P\gamma(2\phi-1)}{1+\gamma^2}\right)$$

where P is the neutron polarization, ϕ the neutron flipping efficiency, F^m_{hkl} and F^n_{hkl} the magnetic and nuclear structure factor respectively and

$$\gamma = \frac{F^m_{hkl}}{F^n_{hkl}}$$

From the above relationship a further advantage of the use of polarized neutrons is apparent. Not only the magnetic scattering is enhanced as given by Figure 4, but in the determination of the polarization ratio the instrumental constant cancels out and it is then possible to determine γ_{hkl}, that is F^m_{hkl}, (one assumes that F^n_{hkl} is known), without any absolute determination of intensity and also without an intensity determination relative to some standard (for instance nickel).

As far as the experimental determination of γ_{hkl} is concerned, it is easy to show the following points [2]:

1. As a monochromator it is convenient to have a crystal with a rather large mosaic spread in order to increase the reflectivity. One could think of several good polarizer monochromators but the ones of high reflectivity would be in general exceedingly expensive [3].

   On the other hand, it can be shown that the flipping efficiency stays practically unchanged for the wavelength spread connected with the common values of mosaic spread and collimation

2. The sensitivity of the method is given in Figure 6 where it is apparent that the highest sensitivity is for $\gamma \sim 1$.
3. In order to decide about the convenience of using the polarization ratio technique it is however necessary to consider also the relative error in γ_{hkl} for a given error affecting the corresponding R_{hkl}. From Figure 7 one sees that this technique provides a very accurate determination of γ_{hkl} in the range 0.3 to 0.9 . For γ_{hkl} approaching either 0 or 1 the error propagation becomes very unfavourable.
4. An accurate check of the main source of error is fundamental for avoiding entirely misleading results. They include counting statistics, extinction effects, depolarization of the incident beam within the sample and on the walls of the experimental set-up, contamination of half-wavelength neutrons in the monochromatic beam, multiple scattering, etc. All these errors affect the polarization ratio to various extent and a careful analysis has to be carried out to eliminate them or at least to reduce them and apply suitable corrections. Although these corrections have become standard, it might however be worthwhile to mention one of the procedures being followed for correcting extinction, which is the main source of error on R.

Usually, for a certain reflection, the structure factors for incident neutrons of opposite polarization are quite different and extinction, being related to the structure factor, affects in a different way the numerator and the denominator of the ratio R_{hkl}. The presence of extinction is checked by measuring the γ-values as a function of the length of the neutron optical path in the sample and by changing the neutron wavelength. Only γ-values affected by extinction for less than 3% are retained and are corrected by extrapolation to zero optical path and zero neutron wavelength. A check is performed by measuring directly the amount of extinction present in some reflections. The measurement is performed according to a method recently used for X-rays by De Marco [4] The intensity variation of the beams diffracted and transmitted by the crystal when rocked through a Bragg position is determined. The peak in the diffracted intensity and the dip in the transmitted intensity must be equal point by point within statistical errors; the fractional

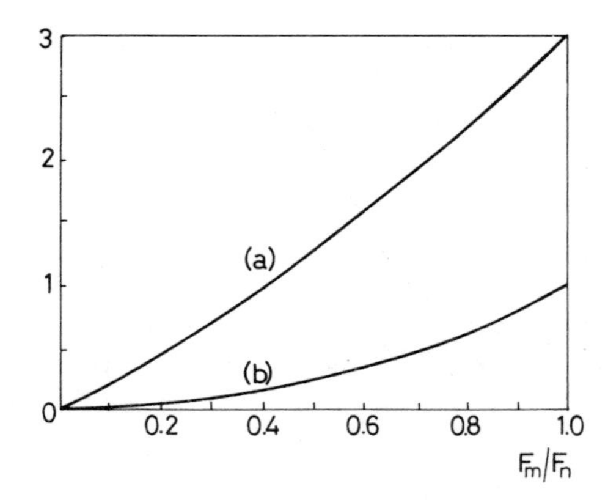

Figure 4 Cross-section for magnetic scattering for polarized and unpolarized neutrons respectively.
(Reproduced by permission of North-Holland Publishing Company)

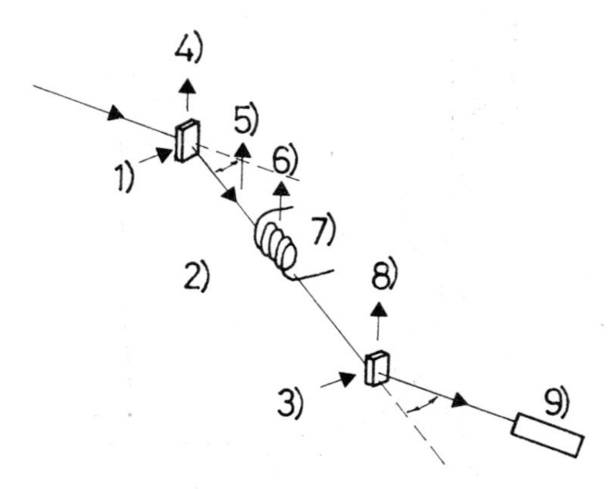

Figure 5 Sketch of the experimental set-up of a polarized neutron spectrometer for magnetization density experiments:
1) polarizing crystal; 2) polarizing inverter ;
3) analysing crystal; 4) polarizing field;
5) collimating field; 6) precessing field;
7) R.F. coil; 8) analysing field; 9) detector;
from R. Nathans et al. (8b).
(Reproduced by permission of Pergamon Press Ltd.)

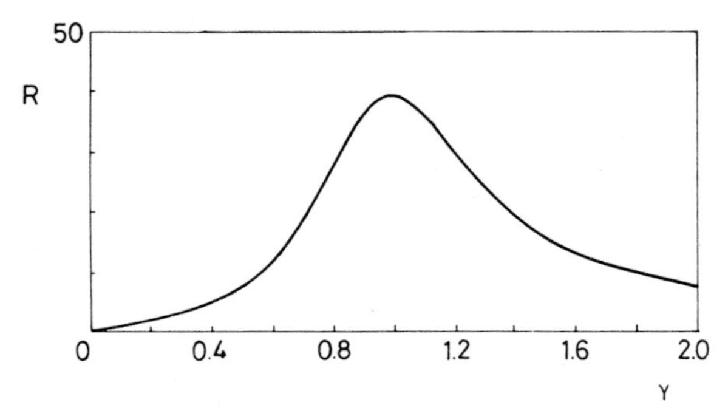

Figure 6 Polarization ratio R as a function of γ for a beam polarization P = 0.99 and a flipping efficiency ϕ = 0.98.

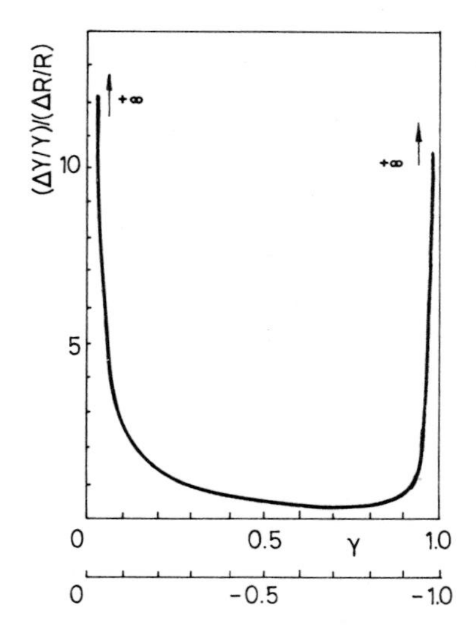

Figure 7 Propagation of the experimental errors on R as a function of γ (from A. Paoletti and F.P. Ricci (2)).

decrease of the transmitted beam gives the amount of extinction.

In particular cases also polarization analysis of scattered neutrons can provide useful information. By this procedure neutron scattering associated with spin flip can be separated from other scattering components. For instance in paramagnetic scattering if the scattering vector is parallel to the neutron polarization, the scattering will be entirely spin flip and in this way is separated from other much stronger scattering components. The spin analyzer consists of a crystal of zero reflectivity for neutrons with spin of a given orientation. The spin analyzer is mounted at the place of the crystal analyzer in a triple axis spectrometer properly equipped with polarizing monochromator, guide fields and spin flipping devices.

Magnetization densities by elastic scattering of unpolarized neutrons

A standard double axis spectrometer is used and the standard precautions are taken to optimize luminosity resolution [5]. In measurements with unpolarized neutrons the problem of performing absolute determinations of magnetic intensities arises and this requires accurate knowledge of the nuclear scattering amplitudes and the temperature factors. Furthermore for ferromagnets the problem of distinguishing the magnetic from the nuclear scattering is also a severe one. In simple antiferromagnets nuclear and magnetic peaks are usually separated.

However in placing the observed intensities on an absolute basis it must be reminded that in practice contamination of nuclear peaks by magnetic scattering and vice versa can be present. Also corrections for second order contamination of the monochromatic beam are much more important than for the polarized neutron method.

Unpolarized neutron experiments are often performed on single crystals for intensity sake, but it must be considered that for accurate measurements, powder samples have the general advantage that errors due to extinction and absorption are smaller.

Magnetization densities by inelastic scattering of polarized neutrons

As already pointed out, experimental determination of magnetic form factors at general positions of the reciprocal space can provide information complementary to the results obtained at reciprocal lattice points.

In a ferromagnetic material, in the expression for the phonon creation cross section, magnetic terms appear, which are elastic in the spin system but inelastic in the nuclei system: this is known as magnetovibrational scattering. A polarized beam technique can be used to obtain the ratio of the magnetic to the nuclear scattering amplitude for phonon scattering as in the case of Bragg scattering, (the dependence on neutron polarization being the same). Measurements can then be taken away from the Bragg peaks.

A triple axis polarized beam spectrometer is used with a polarization independent analyzer on the third arm. The intensities of phonon peaks are measured for neutron polarization parallel and antiparallel to the sample magnetization, the sample being saturated in a direction normal to the scattering vector.

From the ratio of the intensities the magnetic scattering amplitude and therefore the form factors, can be deduced for that particular $\underline{K}$ involved in the scattering experiment. The problem of detecting possible differences between the elastic and inelastic form factors is a very interesting one for the reason already discussed. However preliminary measurements on Ni and Fe do not provide evidence that such differences do exist[6].

Comparison between experimental methods

The polarized neutron method is usually applied to ferromagnets and in some cases to paramagnets if they can be conveniently magnetized. For antiferromagnets the method cannot be used in principle as the coexistance of opposite magnetized sublattices determines a polarization ratio in general equal to unity. In antiferromagnetic domains where magnetic atoms with reversed magnetization are related by a simple lattice translation, for every reflection, either the nuclear or the magnetic structure factors are zero so that $R = 1$. However $R \neq 1$ if:

1) The magnetic atoms with reversed spin are related by one of the symmetry elements of the space group and not by a lattice translation
2) the nuclear and magnetic structure factors are not 90° out of phase, that is when magnetic atoms related by a centre of symmetry have parallel spin
3) one of the antiferromagnetic domains which can be present occupies a predominant fraction of the crystal volume[7].

Therefore in few antiferromagnetic systems, polarized neutrons can provide valuable information on magnetization densities at particular sites of the unit cell.

The use of polarized neutrons is also extremely useful in magnetic compounds where the introduction of some covalency into the predominantly ionic configuration can be expected to modify the distribution of unpaired electrons, mostly in the region between cation and anion. The study of possible magnetization density away from the magnetic ion is practically impossible with unpolarized neutrons because it contributes mostly to the innermost magnetic reflections where departures from spherical symmetry of magnetization are not evident and the contribution from the almost spherical magnetization density localized about each of the magnetic atoms is overwhelming.

Unpolarized neutrons must be used for determining magnetic structure factors in all the systems where the magnetic scattering is insensitive to neutron polarization: i.e. where the polarization ratio R does not deviate from unity as in the majority of antiferromagnets. In these cases magnetic structure factors must be determined by magnetic elastic scattering of unpolarized neutrons. It must be pointed out that unpolarized neutrons have to be used sometimes in ferromagnets as well. From Figure 7 one sees that the propagation on $\gamma_{hk\ell}$ of the experimental errors on R is particularly unfavourable for $\gamma_{hk\ell}$ approaching either 0 or 1. However it is still convenient to use polarized neutrons for measuring the magnetic structure factors if $\gamma_{hk\ell} \to 0$, as the errors involved with an unpolarized-neutron measurement would be larger. For $\gamma_{hk\ell} \to 1$, i.e. for $F^m_{hk\ell} \simeq F^n_{hk\ell}$, unpolarized neutrons give more reliable information also if the polarization ratio method reaches the highest sensitivity (Figure 6). A typical example is provided by the (111) and (200) reflections of f.c.c. cobalt.

Outline of experimental results

Magnetization densities have been measured in several metals, alloys and compounds and a comprehensive review of results is outside the scope of this book. However it must be pointed out that neutron scattering has provided unique and sometimes unexpected information on the distribution of magnetic electrons and has made possible a direct comparison with calculations based on various theoretical methods. In the following section some examples of particularly significant results in the study of magnetization densities in different systems will be given.

Metals and compounds of transition elements

The first measurements made by unpolarized neutrons indicated for experimental magnetic form factors a behaviour in general agreement with free-atom-one-electron calculations but did not permit one to see many details, until in 1959 the first polarized-neutron measurements[8] gave evidence of the departure of 3d electrons from spherical symmetry in cobalt, iron and nickel. That was interpreted in terms of deviation of the T_{2g} and E_g sublevel populations, from the statistical value (60% and 40% respectively).

It is well known that for the sake of simplicity the theoretical calculations assume electron wave functions to be separable into radial and angular parts (potential spherical symmetric). The accuracy of such an approximation might be questioned but it is not easy to do much better. The assumption of separability permits one to obtain an exact solution of hydrogen orbital character for the angular part and to evaluate numerically or to expand in terms of convenient functions the radial part. However, for outer electrons one prefers to consider orbitals which recall the symmetry of the crystalline field. Thus for d electrons in a cubic lattice one finds that suitable combinations of hydrogen orbitals with $\ell = 2$ lead to charge densities with cubic symmetry. One can obtain (I) a combination called E_g which can accommodate two of the five d electrons of each spin with charge density pointing along the six (100) directions and (II) another combination, T_{2g}, which can accommodate the remaining three electrons of each spin with charge density pointing along the eight (111) directions.

The orbital charge densities are respectively

$$E_g = \frac{15}{16\pi}\left[\frac{x^4+y^4+z^4-2x^2y^2-2x^2z^2-2y^2z^2}{(x^2+y^2+z^2)^2} + \frac{1}{3}\right]$$

and

$$T_{2g} = \frac{15}{4\pi}\left[\frac{(xy)^2+(xz)^2+(yz)^2}{(x^2+y^2+z^2)^2}\right]$$

and are shown in Figure 8.

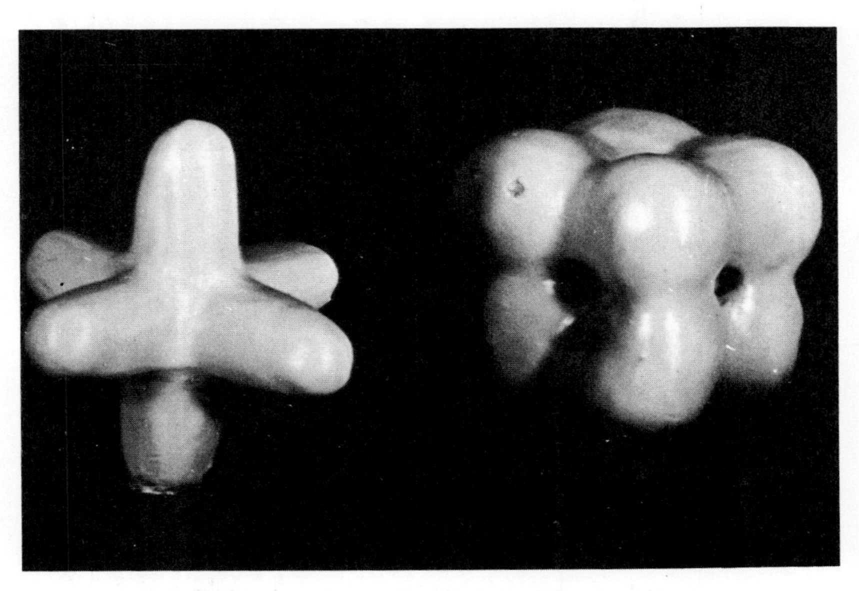

Figure 8 Models of the E_g (left) and T_{2g} (right) orbital charge density. The E_g points toward the six [h00] directions, while the T_{2g} points in the eight [hhh] directions from (R.J. Weiss (B6) p. 64).
(Reproduced by permission of North-Holland Publishing Company)

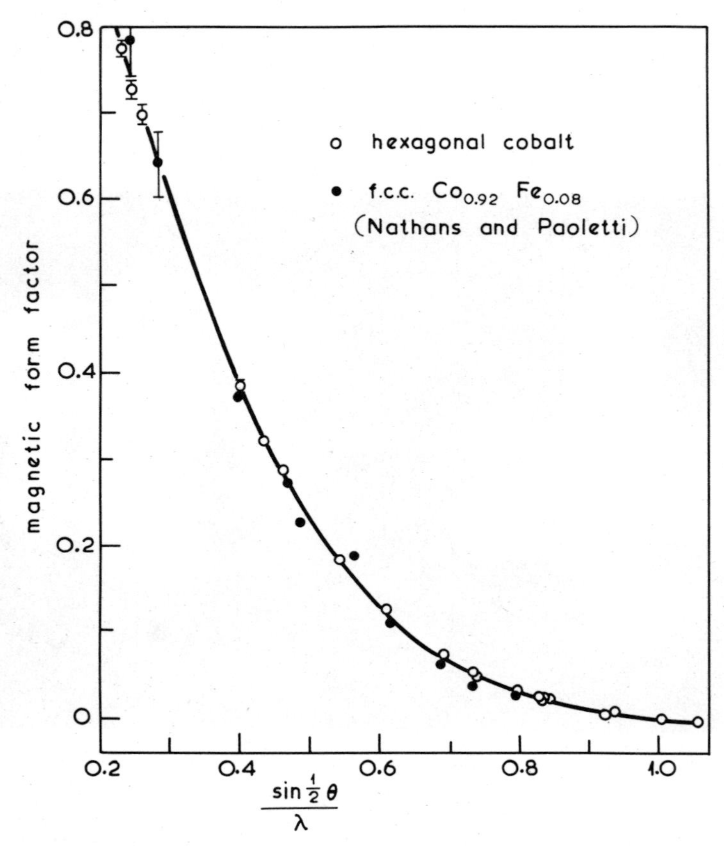

Figure 9 The magnetic form factor of h.c.c. cobalt compared with f.c.c. cobalt. O hexagonal cobalt. (R.M. Moon 1a) O f.c.c. cobalt (R. Nathans and A. Paoletti (8a) (from W. Marshall, S.W. Lovesey: (B4) p. 195).
(Reproduced by permission of Oxford University Press)

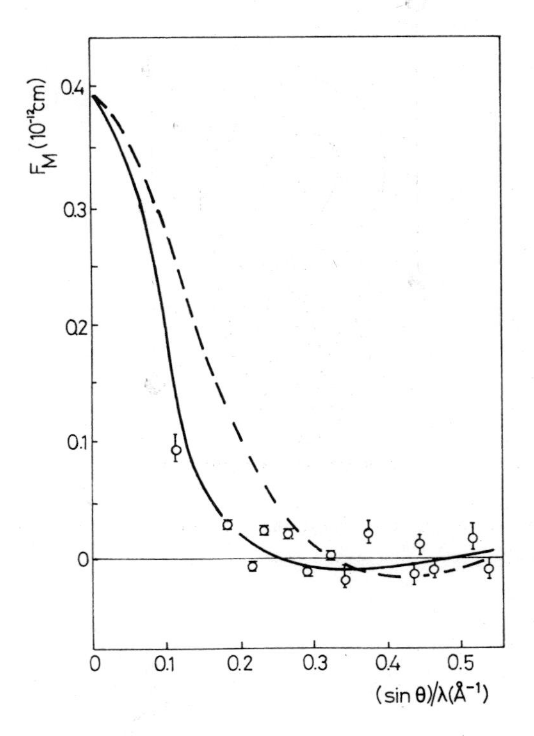

Figure 10 Observed magnetic structure factors of $ZrZn_2$ at 4.2°K: most of the experimental points do not fall on a smooth curve (from S.J. Pickart et al. (48)).

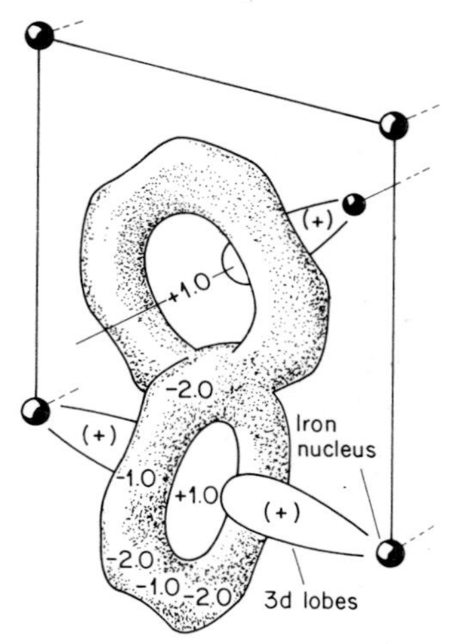

Figure 11 Three dimensional magnetization density in iron (from Shull and Mook (1c)) .

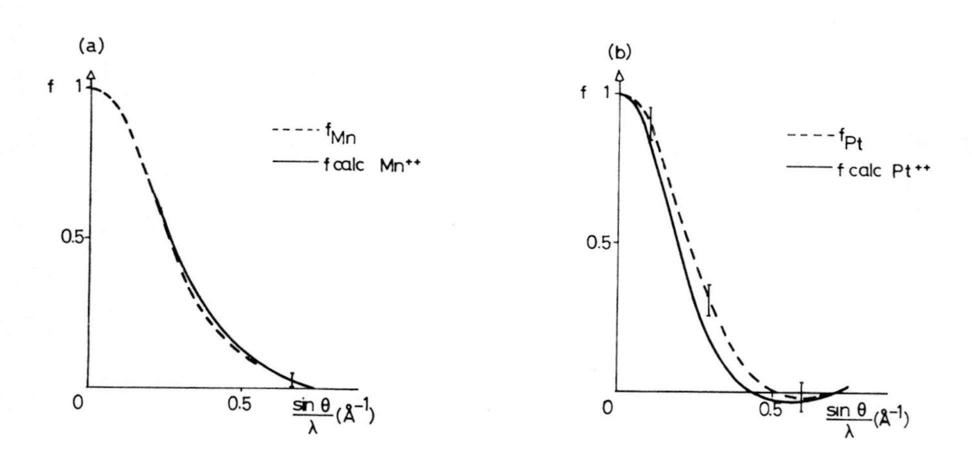

Figure 12 Comparison of the experimental and theoretical form factor for Mn and Pt in $MnPt_3$ (from F. Menzinger et al. (15)).

Now the sum of E_g and T_{2g} densities is spherically symmetric. For unfilled orbitals (as in magnetic elements) one has spherical symmetry if the electron population is 40% E_g and 60% T_{2g} and the form factor is then given by equation (4').

The departure from spherical symmetry in the distribution of magnetic electrons becomes evident on measuring the form factor as a function of the scattering vector $\underline{k}$. In the case of spherical symmetry, the experimental points lie on a smooth curve, equation (4), (Figure 9 circles). If asymmetries are present then on the one hand experimental points deviate from the smooth curve and on the other hand pairs of form factor values relative to $\underline{k}$ vectors of the same magnitude but different orientation provide different results (Figure 10) e.g. reflection pairs such as (511) and (333) or (300) and (221) having the same Bragg angle).

Figure 11 shows the magnetization density map of iron as reported by Shull et al [1,9]. The interatomic magnetization distribution is characterized by a series of interlocking, negative-magnetization rings arranged in a three-dimensional chain structure. Each ring is centred at the midpoint between second-nearest neighbours in the body centred cubic lattice with the plane of the ring being transverse to the bonding line. There is some variation of the negative field with values ranging as high as -2.0 T (20 kOe). As the nuclear position is approached, the very much larger and positive 3d shell magnetization is encountered with fields as large as 50 T (500 kOe). It is estimated from the volume of negative magnetization rings that as much as 35% of the unit cell volume is characterized by a negative field. As far as the interpretation of the iron data is concerned, a first analysis was made by Shull and Yamada [9], who, along with the spin magnetization, made allowance also for the 3d orbital-momentum scattering and for possible form factor contributions from other electron groups.

A negative contribution to magnetization from 4s electrons was then included for explaining experimental data. Negative polarization was also found in h.c.p. cobalt and nickel. In both cases the magnetic scattering amplitude measured with the polarized neutron method, as a function of $\frac{\sin\theta}{\lambda}$, would extrapolate for $\frac{\sin\theta}{\lambda} = 0$ (where the form factor $f = 1$) to a value definitely larger than the macroscopic magnetization. This result is consistent with a different radial distribution of electrons of opposite spin, the negative spin electrons being so spread out that their contribution to form factor is already negligible for very small values of $\frac{\sin\theta}{\lambda}$. For this reason their negative contribution to magnetization cannot be detected by neutron scattering, but does appear in the bulk magnetization which is therefore lowered.

Experimental results in pure 3d metals have stimulated a considerable amount of theoretical work based both on free-atom and band calculations leading to fair agreement with the magnetization densities measured by polarized neutrons[10].

Besides the study of pure metals, polarized neutrons have been extensively used in ferromagnetic alloys. Of course in these cases the interpretation of the data is more difficult as the magnetic structure factors contain the contribution of magnetic moments and form factors of the various components. The situation of course is simpler for ordered alloys, where one has along with the fundamental reflections also the superlattice ones. The scattering amplitudes appear in the structure factor with different coefficients for superlattice and fundamental peaks respectively. For instance in alloys as Fe_3Al, Fe_3Si one has three different kinds of structure factors:

$$F_1 = 2a_B + a_B + a_D \qquad (h+k+l \text{ even}, \ \frac{h+k+l}{2} \text{ even})$$

$$F_2 = a_D + a_B - 2a_A \qquad (h+k+l \text{ even}, \ \frac{h+k+l}{2} \text{ odd})$$

$$F_3 = a_D - a_B \qquad (h+k+l \text{ odd})$$

a_A, a_B, a_D are the scattering amplitudes for the sites A, B, and D. Therefore in these cases we have three sets of equations for determining the μ_A, μ_B, μ_C moments, if we know the form factors. Usually one assumes the spherical part of the form factor to be, to a first approximation, the same as for pure metal. Therefore, by considering those reflections for which the form factor is practically insensitive to the 3d sublevel population it is possible to calculate the moments and once the moments are known, to determine for each reflection the form factor to be interpreted for instance in terms of a one-electron wave function. Alternatively from experimental data one can draw the magnetization density map in which it is possible to associate with each kind of atom, the magnetization density of surrounding area. When the alloy is disordered of course only average information can be obtained and also the attribution of the moments to the components become less certain. The only direct method of investigation in these cases is the neutron incoherent elastic scattering which was primarily used with unpolarized neutrons by Shull and Wilkinson [11] in Fe-Ni and Co-Ni alloys.

However it is particularly difficult to separate the magnetic contribution from the several other components of incoherent scattering. Bragg scattering can also be of some use in disordered alloys in particular cases, for instance when the form factor of one of the two components falls rapidly to zero or when the extrapolation of the magnetic structure factors to a zero value of $\frac{\sin\theta}{\lambda}$ can be obtained with good accuracy. Otherwise it is necessary to perform a multiparameter fitting of the experimental data. The study of ferromagnetic alloys is in general important in order to compare experimental results with band theory calculations, or simply in order to obtain information on nearest-neighbour interactions.

A typical example is provided by the Co-Ni system [12]. The concentration of magnetic moment distribution has been determined by polarized neutron scattering. The presence of negative magnetic moment density has been detected in all alloys and 3d spin part of the magnetic moment has been found to vary linearly with the concentration. The population of E_g and T_{2g} levels could also be obtained at different concentrations, giving information on asphericities in magnetic electron distribution . The experimental data were originally compared with calculations based on a simple rigid band model. Recently the coherent potential approximation [13] has been extended to treat the case of multiple band in ferromagnetic substitutional binary alloys, leading to a very satisfactory agreement with experimental results.

The study of some alloys might also represent the only way to obtain information on 4d and 5d electrons. None of the 4d and 5d transition metals exhibits a magnetically ordered state as an element but some of their alloys show bulk magnetization properties which are, interpreted as an indication of a contribution of 4d and 5d electrons to the magnetization.

Polarized-neutron analysis has made is possible to obtain valuable information on magnet-moment form factors and magnetization densities on several 4d and 5d metals. As an example we report the results of a study on the alloy $MnPt_3$[14,15]. The moments on Mn and Pt sites as determined by polarized neutrons turn out to be $3.72 \pm 0.03\mu_B$ and $0.14 \pm 0.04\mu_B$ respectively and from experimental form factors for manganese and platinum the deviation from spherical symmetry is evident. The experimental spherical form factors for Mn and Pt , are compared with calculations in Figure 12. In Figure 13 the three dimensional magnetization density at the Pt side is given. The negative density at the centre of Pt sites could be accounted for by assuming that the 1.8% Mn atoms present at Pt sites because of the incomplete order, would carry a moment of about the same magnitude but opposite orientation of the ones at Mn sites. Such a hypothesis is consistent with the fact that for Mn-Mn pairs, the interaction is usually antiferromagnetic and indeed any Mn atom on Pt sites has four Mn nearest neighbours. Measurements on Mn-Pt alloys richer in Mn than $MnPt_3$ have provided further support to this interpretation. Magnetization densities have been also studied in compounds,[11] where interesting effects due to covalency have been detected[11]. A typical example is provided by MnF_2 which satisfies the above mentioned conditions for the use of polarized neutrons in antiferromagnets. The presence of covalancy effects in MnF_2 leads to a total spin density that lacks translational symmetry, since the two manganese atoms are not equivalent, and as a consequence magnetic contributions to the pure nuclear reflections appear [7] . The Fourier inversion of experimental data indicated the presence of covalent spin along the line joining a Mn:F pair, while the presence of contours of both signs indicated that spin density may undergo a change in spin (Figure 14).

Recently the polarized neutron method has been applied to the study of ferromagnetic yttrium iron garnet[16], where Fe^{+3} ion are present in octahedral (a) and tetrahedral (d) sites. Four different groups of reflections are present. One of them depends on tetrahedral iron, two on both tetrahedral and octahedral iron, while oxygen atoms only contribute to the last group. The first three groups of reflections were used for obtaining magnetic moments and form factors

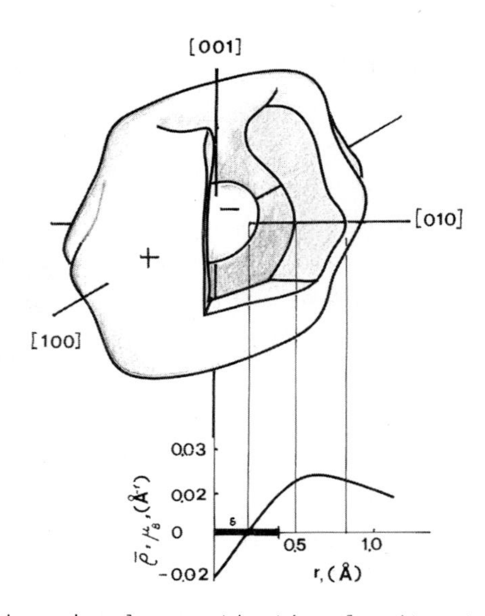

Figure 13 Three-dimensional magnetization density at the Pt site in a $MnPt_3$ alloy with an order parameter S = 0.90 (from A. Paoletti (10a)).
(Reproduced by permission of Società Italiana di Fisica)

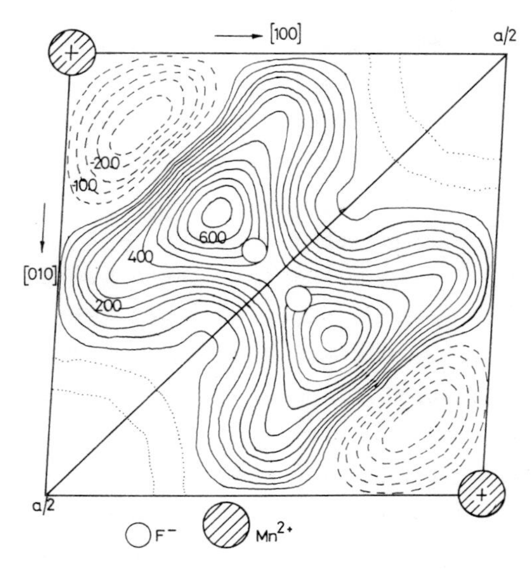

Figure 14 Covalent spin density in MnF_2 projected on the basal plane, obtained from the measured "forbidden" magnetic structure factors; (from R. Nathans et al. (7)).

for octahedral and tetrahedral Fe^{+3} ions. The magnetic moments were calculated directly from experimental magnetic structure factors eq. 2 and turned out to be $-3.70\mu_B$ and $3.75\mu_B$ for Fe^{+3} ions in tetrahedral and octahedral sites respectively, with a magnetization per formula unit of $3.60\mu_B$ in very good agreement with bulk magnetization data. Magnetic form factor for the octahedral site is in good agreement with free ion calculations. For tetrahedral site a considerable deviation is apparent, probably due to covalency effects. Also to covalency has been ascribed the net magnetization density at oxygen sites, obtained from the polarization ratio ($\neq 1$) measured for the class of reflections which depends only on oxygen atoms.

Metals and compounds of Rare Earths

In rare earth metals the orbital contribution to magnetic moments is very important and magnetic anisotropy is usually high. That introduces considerable complications in the study of magnetic form factors and the polarized neutron technique is only partially suitable. As a matter of fact, in the 3d case, where the unpaired spins constitute the predominant part of the magnetic moment, the anisotropy of the spin density is not influenced by application of a magnetic field in the sense that only the spin direction follows the magnetic field but its spatial distribution determined by the crystalline fields does not change. In rare earth metals the presence of an unquenched orbital momentum determines a spin-orbit coupling which is larger than crystalline field effect.

Consequently when a large enough magnetic field is applied to align the moment, as is required for the polarized-beam experiment, this field direction becomes the unique axis of moment distribution, so that the cloud of the moment distribution turns with the field. Therefore we can measure with the polarized-neutron technique only the projection of the moment density on a plane normal to the zone axis along which the magnetic field is applied. This implies that also unpolarized neutrons without applied field must be used in order to get the form factor and, more important, that the two sets of data are not necessarily identical.

For instance, terbium has h.c.p. structure and exhibits helical antiferromagnetism below $230^{\circ}K$ and ferromagnetism below $220^{\circ}K$. The saturation moment is of $9.34\mu_B$ along the easy b-axis and the deviation of this number from the free-ion value of $9\mu_B$ has been attributed to conduction electron polarization. Also in this case, of course, the form factor can be divided into a spherical and an aspherical part: $f(K) = f_s(k) - \Delta a$. The experimental data indicated the presence of a nonlocalized polarization which gave a contribution of $0.48\mu_B$/atom, which has been ascribed to conduction electron polarization and is of the right value for explaining the difference between the total magnetic moment per atom and the spin and orbital contribution of 4f electrons.

The case of gadolinium would be much more simple, as gadolinium is one of the simplest magnetic materials known with a relatively low anisotropy energy, except for the extremely high absorption cross section for thermal neutrons of the naturally occurring element. A polarized neutron investigation has then been carried out on a sample highly enriched in the isotope ^{160}Gd, in order to gain information on magnetization distribution due to 4f electrons and to evaluate the contribution of conduction electrons[17]. The unpolarized beam method was also used in a certain range of $\frac{\sin\theta}{\lambda}$ where $\gamma \simeq 1$ and the precision of the polarized beam method is extremely low as already pointed out. The measured structure factors indicate spherical symmetry in the 4f spin distribution associated however with an appreciable departure from free ion calculation. Moreover experimental data suggest that magnetization density, which in this case is coincident with the unpaired spin density, as in Gd one deals with an S-state, local moment consists of a localized 4f part and a diffuse part, most probably due to unpaired conduction electrons. By separating the two contributions, the distribution of the diffuse component of unpaired spin density could be determined: at each atomic site there is a maximum density parallel to the local moment. Spin density decreases, goes to zero and reverses to a negative value with the distance (Figure 15). The departure of experimental 4f form factor from Hartree-Fock calculation, has been attributed mainly to additional screening of the nuclear charge by conduction electrons. More recent relativistic band calculation seem to agree with experimental data fairly well. Magnetic form factor has been also measured for Gd^{+3} ion in Gd_2O_3 in the paramagnetic state with the polarization-analysis technique, indicating no appreciable difference from the metallic form

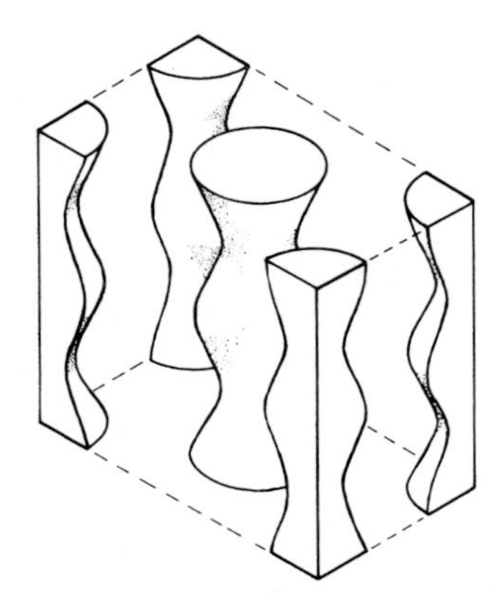

Figure 15 Distribution of the diffuse component of impaired spin density in the unit cell of Gd. The lines show the zero contour; inside the figures so formed the density is negative (from R.M. Moon et al. (17)).

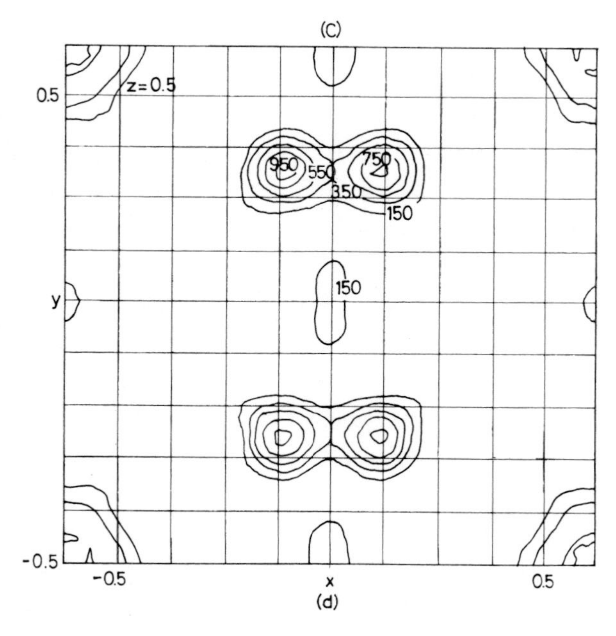

Figure 16 Section at Z = 0.5 of magnetization density in oxygen (from Cox et al. (19a)).

factor[18]. Good agreement with relativistic calculation has been also found.

Spin density by p electrons

Among the few known magnetic material in which the magnetic properties arise from unpaired p electrons, solid oxygen is the simplest one. A detailed polarized neutron investigation has been performed on the paramagnetic cube phase of oxygen in an applied field of 8T (80 kOe), at $46^{o}K$ [(19)].

A major difficulty for the experiment originated from severe extinction which heavily affected the polarization ratios of the strong reflections. Besides peak intensities, integrated intensities were also measured with neutrons in both states of polarization, for several peaks, in order to determine by a least-square procedure the magnetic structure factor, the instrumental scale factor and the extinction coefficient. As the refinement is particularly insensitive to the magnetic structure factor, the extinction coefficient was used to correct the measured polarization ratios.

In this way correct values of the structure factors were obtained to be inserted in a three-dimensional Fourier synthesis providing spin density from p electrons (Figure 16). In oxygen electrons are described by molecular rather than atomic orbitals and the experimental magnetic densities provide valuable information to be compared with theory.

Neutron Investigation of Magnetic Excitations

The excitations of a magnetic system

The energy spectrum of a magnetic system is related to several physical properties both at equilibrium as susceptibility, magnetization, anisotropy, and under dynamic conditions as relaxation times, electrical and thermal conductivity etc. (Figure 17).

Theoretically, the determination of the energy spectrum of a magnetic solid arises from the solution of the problem of free oscillations in a spin system. This problem can be consistently solved in the case of weak oscillations of the system in the framework of the "spin wave" approximation which provides a fair picture of the dynamic behaviour of magnetic crystals at temperature which are low compared to the Curie point. The "Spin wave" theory has been formulated first for magnetic systems described by the "localized electrons" or the Heisenberg model and has been extended to include itinerant electrons[20].

Let us consider a system of spins coupled by ferromagnetic exchange interaction and situated on identical lattice sites $\underline{r}_i$, at absolute zero. A small magnetic field applied along a particular direction (say the z axis) provides an axis of quantization. The ground state of such a system has all the spins parallel to the applied field. Now, if one of the spins is reversed (for instance by raising the temperature), the same exchange forces between neighbours cause a propagation of the reversed spin through the crystal: this is called a spin wave. There is an analogy with the propagation of elastic waves in a lattice , caused by elastic interaction between atoms. Also spin waves can be considered quasi particles, magnons corresponding to the phonons of lattice dynamics.

The problem of one reversed spin in a ferromagnet can be exactly solved by quantum mechanics[21] leading to the relationship that between energy E and wave vector $\underline{q}$.

$$E(\underline{q}) = 2S\,[J(0) - J(\underline{q})] + g\mu_B H \qquad (5)$$

where S is the spin quantum number,

$$J(\underline{q}) = \sum_r J(\underline{r})\exp(i\underline{q}.\underline{r})$$

($J(O)$ being the value of $J(\underline{q})$, for $\underline{q} = O$); $\underline{r}$ is the vector from the origin to the different sites of the lattice.

By considering only nearest neighbour interaction and expresssing the energy in terms of spin wave frequency

$$\hbar\,\omega(\underline{q}) = 2z\,J_e S \left[1 - \frac{1}{z} \sum_{\underline{r}} \exp i\underline{q}\cdot\underline{r} \right] + C \qquad (5')$$

where z is the number of nearest neighbours, J_e is the exchange parameter among them, which is assumed to be isotropic, $\underline{r}$ is the vector from a given atom to one of its nearest neighbours (the sum being over the nearest neighbours) and C is related to the anisotropy and the applied magnetic field. For a cubic lattice (simple, body centred or face centered) if we indicate with a the lattice spacing, equation (5') becomes

$$\hbar\omega = 2S\,J_e\,a^2q^2 + g\mu_B H \qquad (6)$$

or

$$E = Dq^2 + E_o \qquad (6')$$

where D takes the name of stiffness constant by analogy with the mechanical case. Equation (6') can be considered valid at low temperature where only spin waves of long wavelength (small q) are excited. Otherwise terms of higher order in q must be considered and equation (6') becomes

$$\hbar\omega = Dq^2\,(1 - \beta q^2 + \gamma q^4 + \ldots.) \qquad (6'')$$

In Figure 18 the comparison between dispersion curve for nearest neighbour interaction and the small q or quadratic approximation, is given, Magnons like phonons obey the Bose-Einstein statistics. Is is then possible to determine the magnon population at a given temperature and to calculate thermodynamic quantities by inserting (6) in the statistics. The entropy S, specific heat C and magnetization M can be determined. In this way the well known $T^{3/2}$ dependence of magnetization, the famous Bloch result, is obtained.

For magnetic systems more complicated than the exchange coupled ferromagnet, the ground state is not known exactly. For instance the simple picture of an antiferromagnet with the spins of two different sublattices pointing in opposite directions does not represent a state which can be obtained by quantum mechanical calculations, for a system with antiferromagnetic coupling. By perturbation theory however an approximate ground state can be estimated and by assuming a two sublattice model, the magnon spectrum

$$E(\underline{q}) = S\sqrt{z^2 - \left(\sum_{\rho} \cos q.\rho\right)^2} \qquad (7)$$

has been obtained [B5], which is linear in $\underline{q}$ for small q values. In the case of ferrimagnets, if we indicate the spins on the two non-equivalent sublattices by $S_A = (1 + \alpha)S$ and $S_B = (1 - \alpha)S$ where α is a parameter, we have $S_A \neq S_B$, except for $\alpha = 0$ in which case the system is a simple antiferromagnet. With the same procedure above mentioned [B5] one obtains, in the long wavelength approximation, a magnon energy proportional to q^2, which for $\alpha = 0$, reduces to the typical magnon spectrum of antiferromagnets, linear in q.

Equations (6) and (7) represent the so called acoustical branches of the spin wave spectrum which, apart from a small term related to the anisotropy and the applied field, gives zero energy for $\underline{q} = 0$. Also dipolar interaction gives a small contribution at small $\underline{q}$. In more complicated systems where several magnetic interactions are present, one expects[23] besides the acoustical one, also optic branches with finite energies at $\underline{q} = 0$ (Figure 19). The dispersion law is determined in any case by the exchange interactions. Consequently its experimental investigation is extremely valuable for an understanding of the magnetic properties. Neutron scattering provides the most powerful technique for measuring the spin wave spectrum in a significant energy and momentum range. That is due, as already recalled in the introduction, to the particular properties of neutron which, in scattering experiments, can exchange with the magnetic system energy and momentum for an amount which is adequate for the creation or the annihilation of magnons and at the same time can be easily detected by current experimental techniques. The inelastic scattering of neutrons for single magnon creation or annihilation is subject to the laws of conservation of momentum

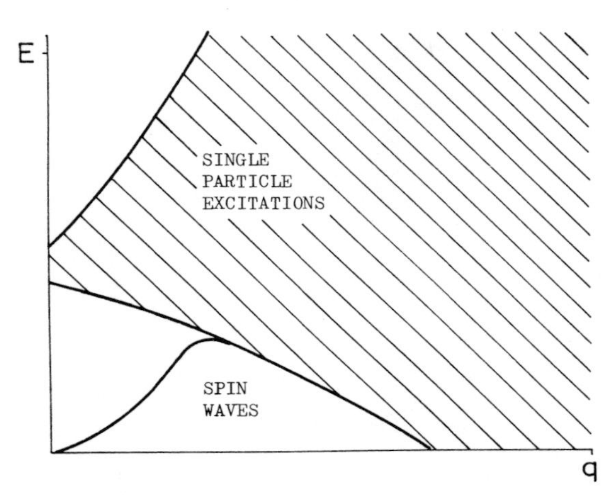

Figure 17 Allowed values of E and q for single particle and collective excitations (from Bjerrum Møller (50) p. 5). (Reproduced by permission of International Atomic Agency)

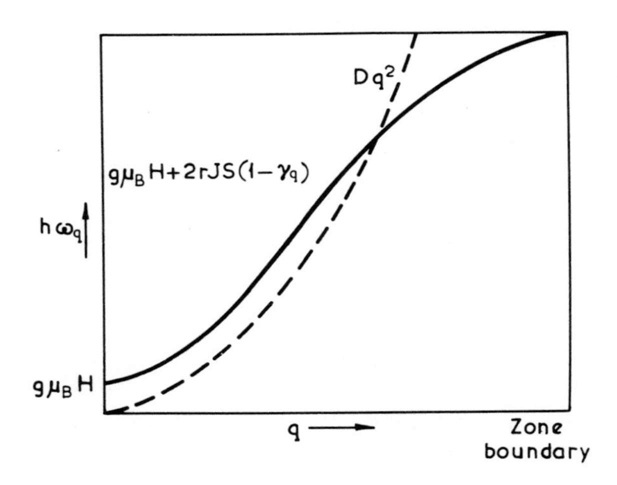

Figure 18 Comparison between the dispersion curve calculated for nearest neighbours interaction and the small q (or quadratic) approximation (from W. Marshall and S.W. Lovesey: (B4) p. 257).
(Reproduced by permission of Oxford University Press)

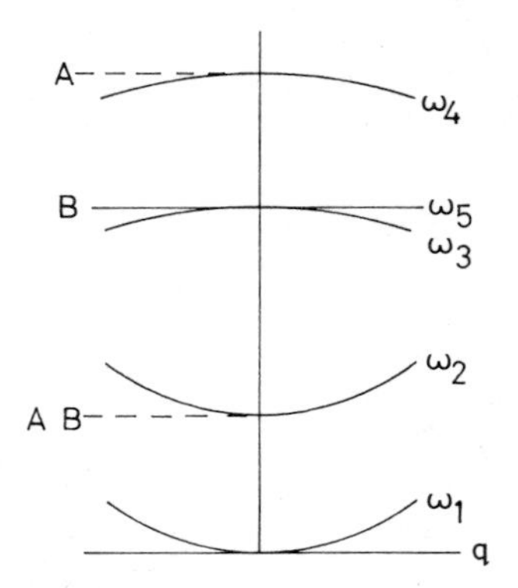

Figure 19 Spin waves dispersion curves in Fe_3O_4 as calculated by T.A. Kaplan (23).

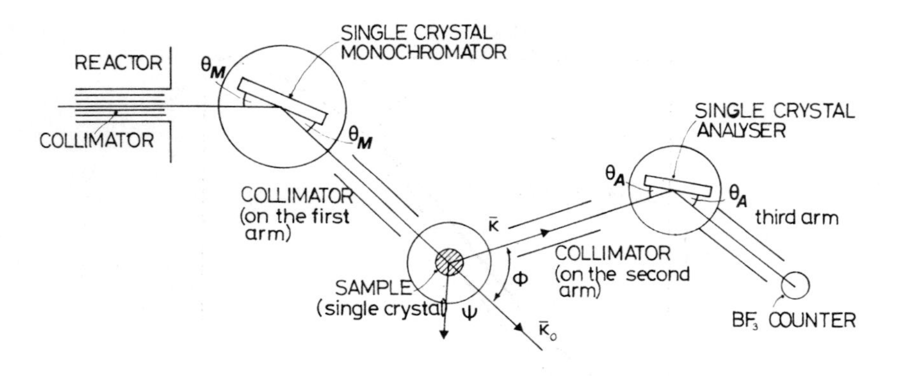

Figure 20 Sketch of a triple axis spectrometer (from B. Buras, (22a)). (Reproduced by permission of International Atomic Agency)

$$\underline{K}_i - \underline{K}_f = 2\pi\underline{\tau} \pm \underline{q} \tag{8}$$

and energy

$$E_i - E_f = \frac{\hbar^2 K_i^2}{2m} - \frac{\hbar^2 K_f^2}{2m} = \pm \hbar\omega(\underline{q}) \tag{9}$$

where $\underline{K}_l$, E_i and $\underline{K}_f$, E_f are the wave vector and the energy for incident and inelastic scattered neutrons respectively, $2\pi\underline{\tau}$ is a reciprocal lattice vector and $\underline{q}$ and $h\omega(\underline{q})$ are the wave vector and the energy of the magnon which has been created or annihilated by the scattering process.

Now equations (8) and (9) imply that in magnetic material a magnon of energy ω can be created by neutron scattering, only if (a) the neutron has an energy higher than $h\omega$, i.e. $K_i - K_f > 0$, and (b) the direction of the incident and scattered neutron ($\underline{K}_i$ and $\underline{K}_f$) are such that a momentum $h\underline{q}$ can be provided , which satisfies the dispersion relation $\omega = \omega(\underline{q})$ of magnons for that material. These are necessary but by no means sufficient conditions for measuring dispersion curves of magnetic excitations. One can recall that inelastic scattering of neutrons can create and annihilate phonons as well and often with higher cross section than magnon, so that the experimental arrangements must permit to discriminate between magnon and phonon inelastic scattering. Moreover one can perform the experiment only in regions of the reciprocal lattice where the contribution of Bragg scattering is very small and on the contrary the magnetic form factor is still appreciable.

Experimental methods

The experimental techniques for determining magnon dispersion curves, which have so far been used, are mainly four. Two of them 1) Spectrometry by triple axis spectrometer, 2) Time of flight method; require a direct energy analysis of the scattered neutrons while the remaining ones: 3) Diffraction method and 4) Small angle scattering method; can be considered indirect procedures as the energy-momentum relationship is not directly measured but rather evaluated by determination of geometric properties of the scattering surface. In the following, a quick review of these experimental techniques is given.

1) Spectrometry by triple axis spectrometer

A sketch of a triple axis spectrometer is given in Figure 20. A white beam of neutrons emerges through a collimator from a reactor hole and a particular wavelength λ_i is selected by a monochromator located on the first axis of the spectrometer. The monochromatic beam of neutrons impinges along a particular direction ($\underline{K}_i$ is then fully determined) on the monocrystalline sample to be studied, which is located on the second axis. The neutrons scattered in a given direction (which determines the orientation of $\underline{K}_f$) are then analyzed by a second monochromator placed on the third axis and the intensity as a function of energy is detected by the counter located on a rotating arm. In this way the intensity for every possible $\underline{K}$, of the same orientation as determined by the orientation of the scattered beam, can be recorded. The orientation of the sample determines $2\pi\underline{\tau}$, so that ω and $\underline{q}$ can be evaluated by equations (8) and (9), all the other quantities being determined. In order to change $\underline{K}_i$, $\underline{K}_f$ and $\underline{\tau}$ the angles θ_M, ψ, ϕ, and θ_A of Figure 20 have to be changed. By repeating the experiment for various $\underline{K}_i$, $\underline{K}_f$ and $\underline{\tau}$ and recording the intensity of neutrons inelastically scattered by magnons i.e. neutrons which satisfy equations (8) and (9), in principle the complete dispersion curve $\omega = \omega(\underline{q})$ could be experimentally determined. It is obvious however that such a procedure in practice would turn out to be extremely time wasting and practically meaningless as the counter would detect together with the neutrons scattered by magnons, those inelastically scattered by phonons, and elastically Bragg scattered plus the background due to both nuclear and magnetic incoherent scattering.

It is then necessary to devise particular procedures to single out the magnon scattered neutrons and to obtain in the quickest possible way the dispersion curve. The most used procedures are the so called "constant $\underline{q}$" and "constant E" scanning.

The "constant $\underline{q}$" method consists in performing measurements of scattered neutrons intensity by steps, changing each time the spectrometer parameters ψ, ϕ, and θ_A in such a way that $\underline{q}$ appearing in (8) stays constant. In other words one looks only for those neutrons which in the scattering process could have exchanged the same momentum, determined by $\underline{q}$. Now from Figure 19 one sees that for a certain $\underline{q}$ value magnons exist only with energies $h\omega$ given by the interception of the vertical line $\underline{q}$ = const. with the dispersion curves, which means that by plotting the neutron intensity as detected by the counter as a function of ω at constant $\underline{q}$, one will see intensity peaks for those ω values which lay on the dispersion curves for that particular $\underline{q}$. The scan is then repeated for a different $\underline{q}$ and the complete dispersion law can then be determined. The value of the parameters are usually changed automatically according to a preset-sequence determined by computer calculations. Analogous procedure is followed in the "constant E" method. In this case only those neutrons are detected which in the scattering process could have exchanged the same energy $h\omega$. In the ωm $\underline{q}$ plot of Figure 19 the "constant $\underline{q}$" method and the "constant E" method correspond to vertical and horizontal scan respectively. The behaviour of the curve ω = ($\underline{q}$) determines which one of the two methods is to be preferred. Further details can be found in review articles by Brockhouse [24] and by P.K. Iyengar (in B2 p97).

In the collection of experimental data the spectrometer resolution function is extremely important. The beam collimation, mosaic spread of the crystals and geometrical parameters of the spectrometer, determine the resolution function which must be known both for a correct interpretation of the experimental data and the optimization of the experimental set up (P.K. Iyengar in B2).

An alternative model of a triple axis spectrometer is shown in Figure 21. A double monochromator is employed in which the beam is reflected in succession from two matched crystals in parallel positions. With this set up the monochromatic beam is parallel to the white beam emerging from the reactor independently of the reflection angle. Consequently the energy of incident neutrons can be changed continuously without moving the specimen[26]. The intensity of the monochromatic beam is lower than in the conventional spectrometers, but there is an appreciable reduction both in the wavelength spread and in the background, so that the signal to noise ratio improves.

In the determination of spin wave spectrum the problem also arises of separating the magnon scattering from other inelastic components such as magneto vibrational scattering. In some cases, it is possible to take advantage of an appreciable difference between the Curie to Néel temperature and the Debye temperature. The intensity peaks of the neutrons inelastically scattered by the excitations corresponding to the higher temperature will appear sharper. In any case the best method for separating magnon and phonon scattering is based on the application of a magnetic field at an angle α with the scattering vector. The nuclear scattering is obviously unaffected by the magnetic field while the magneto-vibrational scattering varies as $\sin^2\alpha$. On the contrary the magnon scattering cross section is proportional to $(1 + \cos^2\alpha)$. Therefore the different intensity variation by the application of a magnetic field permits the identification of the origin of an intensity peak. A clear example is given in Figure 22 which refers to $Co_{.92}Fe_{.08}$. It must be pointed out that in general the triple axis spectrometer is still the most reliable instrument for determining momentum and energy of the scattered neutrons. In the high flux reactors, the largest fraction of experimental work in magnetic excitations is carried out by spectrometers basically designed according to the above mentioned principles. Other technical developments include high reflectivity monochromators, focusing and on-line control.[27] The picture of a three axis spectrometer is given in Figure 23.

2) Time of Flight Spectrometry

The energy of a neutron in a triple axis spectrometer is measured by using in general a single crystal as analyzer. In the time of flight method it is determined the time t, a neutron needs to run a fixed distance ℓ. If λ is the wavelength, m the neutron mass and h Plank's constant

$$\lambda = \frac{h}{m\ell} t \qquad (10)$$

the distance ℓ is also called the flight-path and the time t is the time of flight. Neutrons of different energies have different velocities i.e. different times of flight for a fixed path.

In principle the time of flight method consists in determining by a multichannel time analyzer the intensity of the neutron groups which cover the flight-path in a fixed time interval $t_1 + \Delta t_2$, after a neutron burst has been generated by a chopper at $t = 0$. Usually the chopper is a rotating mechanical device (a disc or a cylinder) made by a material with high absorption cross section for thermal

Figure 21 Sketch of a triple axis spectrometer (from B.N. Brockhouse et al. (26)) .

(Reproduced by permission of International Atomic Agency)

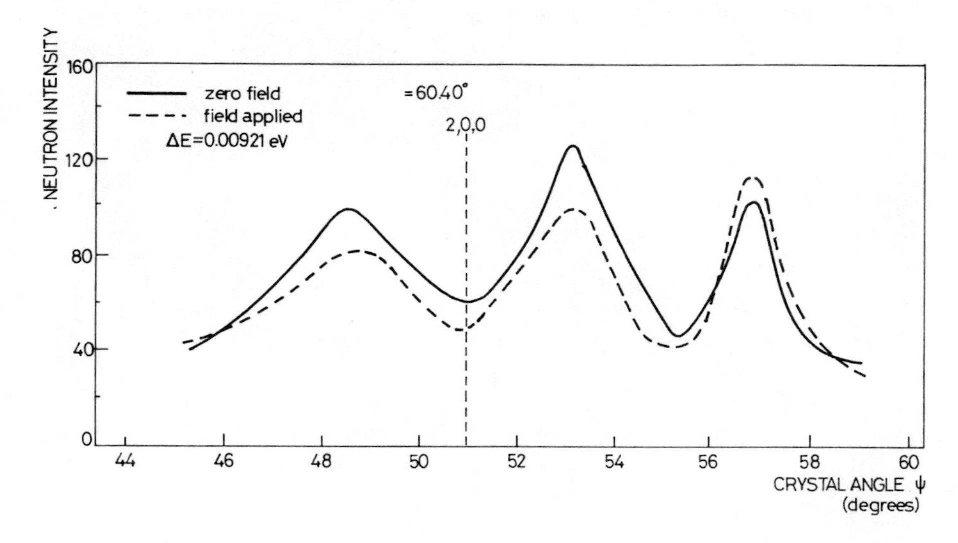

Figure 22 Identification of magnon and phonon peak in $Co_{0.92}Fe_{0.08}$ by their different response to a magnetic field (from Sinclair and Brockhouse (25)).

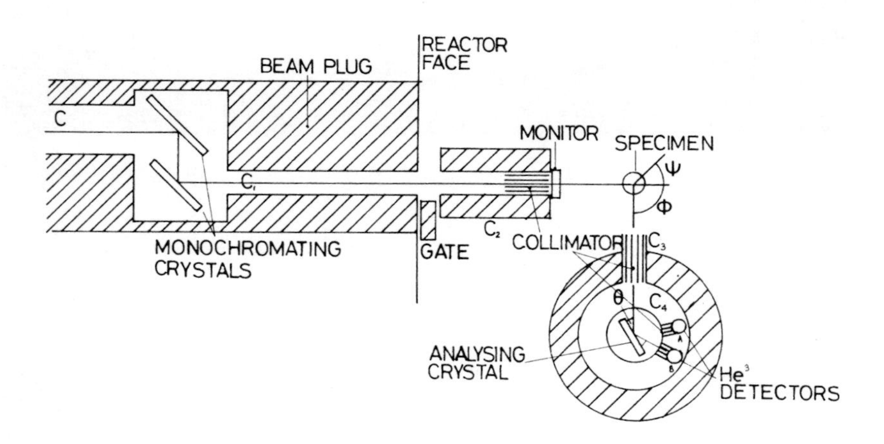

Figure 23 Three axis spectrometer of CASACCIA CENTER of CNEN Rome.

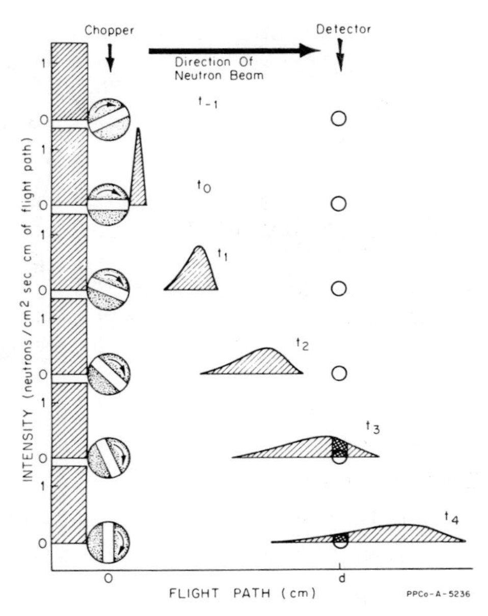

Figure 24 Schematic representation of a time of flight experiment (from R.M. Brugger in B2 pp 55).
(Reproduced by permission of Academic Press Inc. (London) Limited).

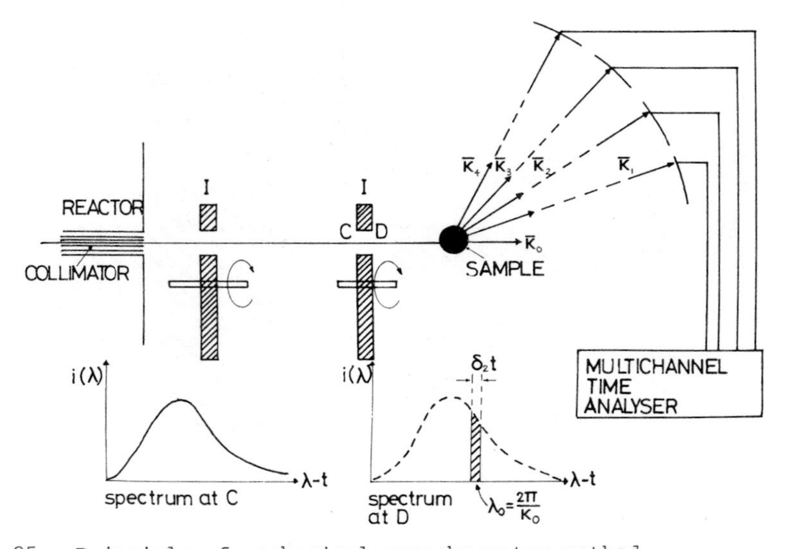

Figure 25 Principle of mechanical monochromator method (from B. Buras: (22b)).
(Reproduced by permission of International Atomic Agency)

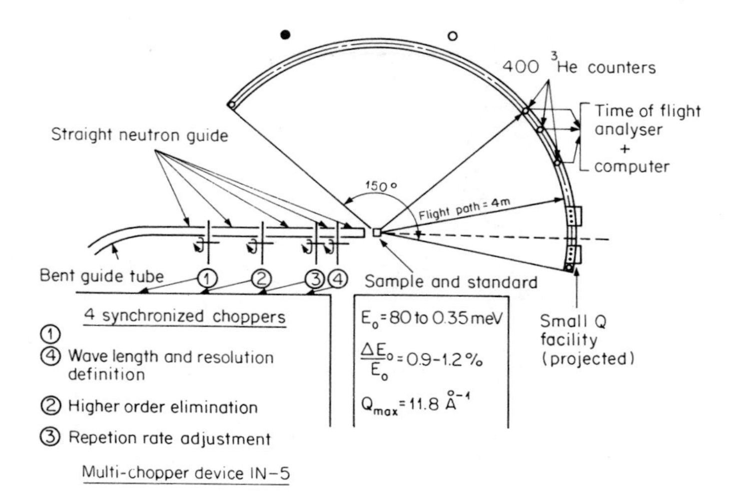

Figure 26 Layout of a multichopper device (from: Neutron beam facilities at the H.F.R. Grenoble January 1974). (Reproduced by permission of Institut Max Von Laue-Paul Langevin)

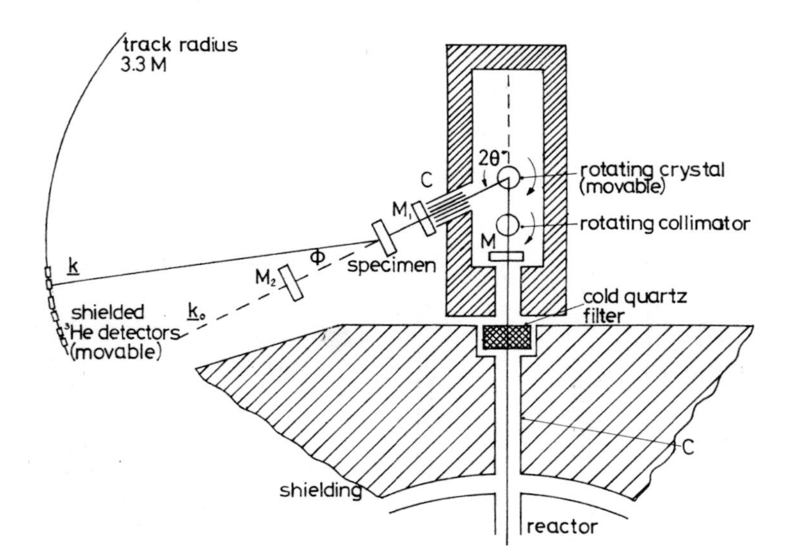

Figure 27 Layout of a rotating crystal spectrometer (from A.D.B. Woods et al.: (34)). (Reproduced by permission of International Atomic Agency)

neutrons. Properly shaped slits are cut through, in order to provide burst of neutrons for a certain angular range of each revolution. The rotation axis can be either perpendicular or parallel to the axis of the incident beam from the neutron source. It must be recalled however that neutron bursts can also be generated by discontinuous neutron sources as pulsed reactors and pulsed accelerators.

A schematic representation of a time of flight experiment is given in Figure 24. Such an experiment allows an energy analysis of a polichromatic beam; it can then be used in inelastic scattering studies to determine the energy distribution of the scattered neutrons after a monochromatic burst has hit the sample. In order to produce bursts of neutrons which in first approximation can be considered monochromatic, several methods have been developed.

a) In the mechanical monochromator method, (Figure 25); this is achieved by combining two choppers placed several meters apart which rotate at the same speed but with a phase difference. A polychromatic neutron bursts passes through the first chopper while the second passes only neutrons having energies in a narrow band with the mean value determined by the distance and the phase difference between the two main ones in order to eliminate "higher order" contamination and to prevent frame overlap. A sketch of recently built multichopper is given in Figure 26.

b) In the rotating crystal method monochromatic neutron pulses can be obtained by rotating a single crystal in a polichromatic beam in such a way that only for very short time intervals the Bragg condition for a particular reflection is fulfilled. In order to prevent significant reflection from unwanted planes and eliminate "higher order" contamination, a chopper or another monochromator can be placed in front of the first monochromator, both devices rotating in phase (Figure 27).

It must be pointed out that the resolution in a time of flight experiment is determined by the transmission function if mechanical choppers are used, while for rotating crystal spectrometers not only the mosaic spread must be considered but also time focusing of neutrons due to Doppler effect[28]. Either in the mechanical monochromator and in the rotating crystal method several neutron counters can be located at the end of the flight path at various angles for simultaneous counting of neutrons inelastically scattered in different directions. In this way neutron groups for several values of ω and $\underline{K}_f$ can be detected. Amongst the techniques based on the time of flight are the correlation-type time of flight spectrometer[29] the statistical chopper[30], the electronic chopper for polarized neutrons[31] etc., which improve the experimental efficiency in particular cases.[27]

3) The Diffraction Method

The determinations of the spin wave dispersion curve for low q values have been performed by the so called diffraction method which provides information in particular cases without requiring energy analysis. This method has been used mostly by experimental groups working at medium flux reactors and has provided valuable information, but the experimental data must be taken with some amount of care in order to avoid apparent inconsistency with the results obtained by other procedures.

The magnon dispersion relation may be determined in some cases[32] by simply measuring the width of the neutron scattering surfaces, i.e. the surfaces in reciprocal space which are the locus of points where the energy and momentum conservation laws are obeyed. It was shown that in the case of a quadratic dispersion relation, the scattering surface is a sphere, while the presence of higher than quadratic terms leads still to a closed scattering surface of quasi-spherical shape only if the quadratic term is the leading one[33]. The scattering process in the reciprocal space is indicated in Figure 28. The crystal mis-set from the Bragg position ψ determines the energy while the angular width of the scattering surface Γ, determines the momentum which are exchanged by incident neutrons. In other words a couple of $\hbar\omega$, q values corresponds to a determined couple of ψ and Γ. Consequently the dispersion curve can be obtained by measuring the width of the scattering surfaces for various crystal mis-sets. If the simple relationship in equation (6') is valid, a linear dependence of $\sin^2(\Gamma/2)$ on $(K_i/K_\ell)^2$ exists ($\underline{K}_\ell = \underline{K}_i + 2\pi\underline{\tau}$) with slope proportional to 1/D. If equation (6') is not valid but the scattering surfaces are still closed, the relationship between $\sin^2(\Gamma/2)$ and (K_i/K_ℓ) is not linear any more and the departure from linearity depends on the higher order terms in q present in equation (6").

Practically the experiment is performed scanning with a counter of limited horizontal acceptance through the scattering surface. If all the neutrons scattered in a plane orthogonal to the plane of the figure are collected, the intensities are proportional to scattering cross section for a "one magnon" process, integrated over a slice of the scattering surface cut by vertical planes. Now it can be shown (W.M. Lomer and G. Low in B6 p 38) that for $\underline{q}$ small compared to $2\pi\underline{\tau}$ and assuming the magnon population and the magnetic form factor to be constant over all the scattering surface, one obtains an intensity profile with a rather flat top and sharp cut offs on both sides. In this way the width of the scattering surface can be determined. The use of polarized neutrons increases the sensitivity of the method. By scanning through the scattering surface with the neutrons in the two polarization states,the contribution of non magnetic scattering can be eliminated and the accuracy in the width determination improved. The diffraction method has been widely used mostly in ferromagnetic systems, where full advantage of polarized neutrons could be taken.

In general the agreement with the results obtained with other techniques was fair. However inconsistencies were noticed in some cases, caused by the fact that quite often the correct intensity profile deviates appreciably from a rectangle[35], because of the variation along the scattering surface of both the neutron population and the magnetic form factor. A typical example of experimental data is given in Figure 29.

4) Small angle scattering

In the diffraction method the scattering is observed at small angular mis-sets from the points of the reciprocal lattice corresponding to ($hk\ell$) planes. If one considers the (000) point of the reciprocal lattice the scattering at small angles from the forward beam must be obtained. As $\underline{\tau} = 0$ the scattering surfaces are independent of orientation of the sample: the only point for $(K_i/K_\ell)^2 = 1$ is determined and the width of the scattering surface is then proportional to 1/D and independent of the energy of incident neutrons. For most ferromagnets $\frac{\Gamma}{2}$ is of the order of 1 degree, therefore measurements at very small angle with extremely good collimation for eliminating the primary beam from the reactor, have to be performed. The nuclear scattering can be accounted for performing experiments with an applied magnetic field parallel and perpendicular to the scattering vector respectively.

5) Comparison between experimental methods

A proper choice between the above mentioned techniques to be used in a magnetic excitation study depends largely on the particular experiment to be done. In a time of flight spectrometer, in order to obtain acceptable performances, the beam must be pulsed in such a way that only a very small fraction ($\simeq$1%) of the neutrons of the required energy available in the reactor are used. On the other side scattered neutrons in a wide energy and angular range can be simultaneously detected, which is an obvious advantage when information for continuous values of ω and $\underline{q}$ is asked for. The triple axis spectrometer cannot in general observe more than one energy at one scattering angle at each time, but it is possible by relatively simple control to limit observations to few significant values of exchanged energies and momentum (E constant and $\underline{q}$ constant methods). Furthermore in a continuous reactor the spectrometer has the obvious advantage to utilize all the available neutrons. The rotating crystal spectrometer has over the multichopper system the advantage of mechanical simplicity, higher reliability and lower cost but, because of the limited reflectivity of the crystal, suffers loss of intensity which however is partially balanced by the lower background (the specimen is not in line with the main beam from the reactor). In the case of pulsed neutron sources (pulsed reactors, accelerators), of course only pulsed spectrometers are used.

The diffraction method can be very useful for gaining information on the general behaviour of the dispersion curve in particularly simple cases. The circumstances that the energy-wave vector relationship is only indirectly deduced, makes extremely critical the treatment of the data. As it has been pointed out, some published data, obtained by this technique, might undergo some revision[36].

The small angle scattering has the severe limitation of the possibility of measuring only D and not the dispersion curve. However the advantage of using a white neutron beam and polycrystalline sample makes this technique particular suitable in some cases (e.g. study of the temperature dependence, or concentration dependence of

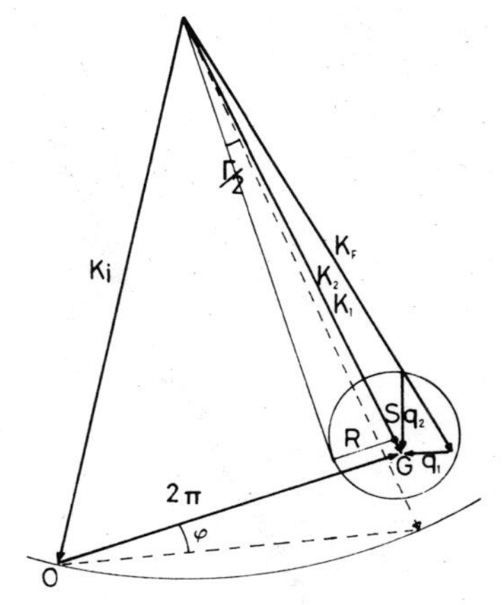

Figure 28 Neutron scattering process by spin waves (from B. Antonini et al. (35)).
(Reproduced by permission of North-Holland Publishing Company)

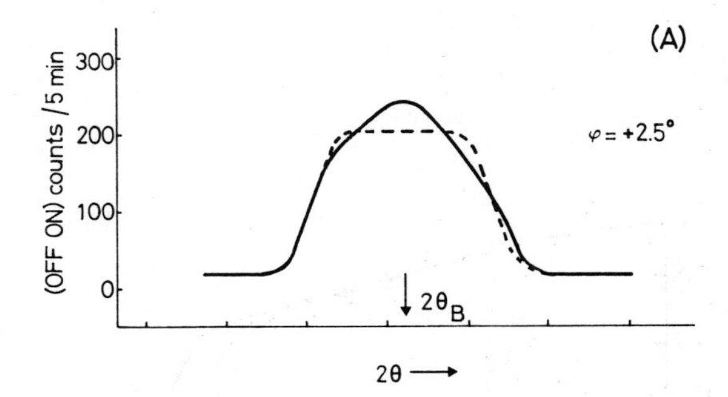

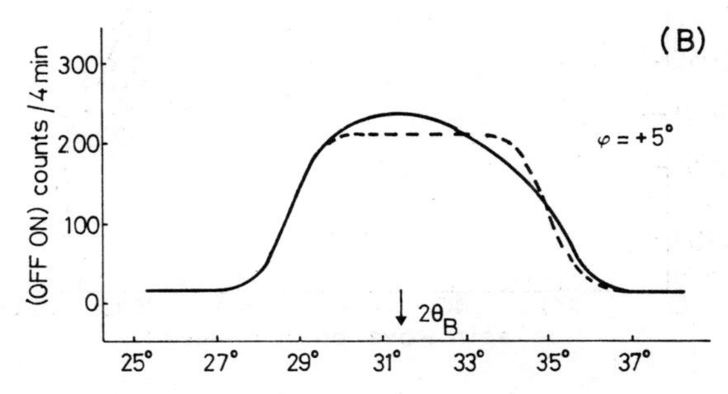

Figure 29 Magnon peak profiles by diffraction method (from B. Antonini et al. (35)).
(Reproduced by permission of North-Holland Publishing Company)

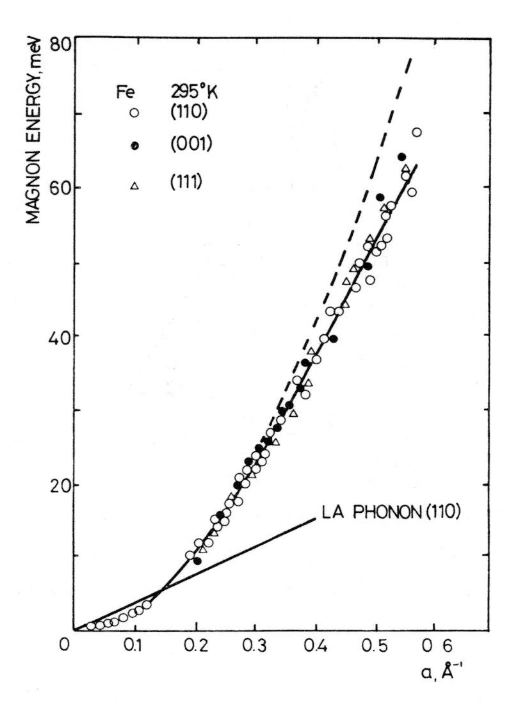

Figure 30 Dispersion curve for Fe (from G. Shirane et al. (51)).

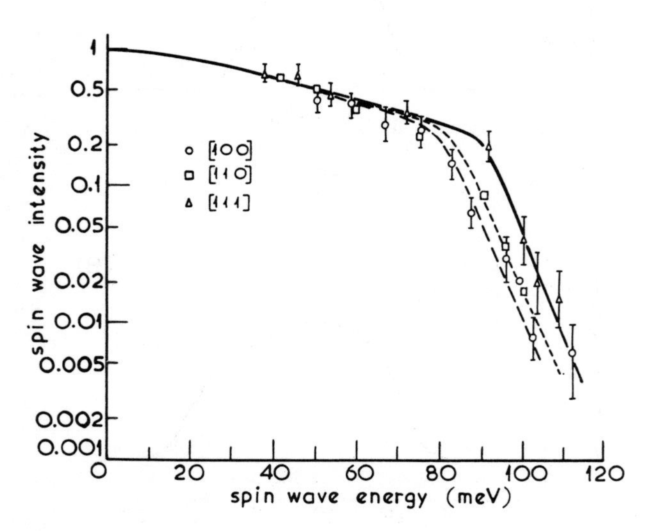

Figure 31 Spin wave intensity as a function of energy in Fe (from H.A. Mook and R.M. Nicklow (38)).
(Reproduced by permission of Journal de Physique)

the stiffness constant).

Outline of experimental results

Spin waves in several magnetic systems have been studied by inelastic scattering, and a full review of the results is beyond the scope of this book. Typical examples only will be discussed. It might be useful to point out that dispersion curves could be often determined only for a limited range of the magnon wave vector q, within the first Brillouin zone. This limitation affects the experimental determination of exchange integrals, and makes sometimes difficult the comparison between results obtained with different techniques.

1) Metals and compounds of transition elements

The study of spin waves in transition metals is made particularly difficult by the high values for the energies involved , and in general the dispersion relation only for rather small wave-vectors has been measured. Both triple axis spectrometry and diffraction method have been used for determining dispersion curves for Fe, Co, Ni and several of their alloys. Iron exhibits an isotropic dispersion curve (Figure 30), furthermore an appreciable deviation from the behaviour expected according to the Heisenberg model was also observed It may be noticed that the iron data were obtained by using the E constant method because of the very high slope of the spin wave dispersion curve. The parameters D and β of equation (6') have been evaluated in several cases. Quite often experimental data would not be explained in the framework of a localized electron model also if long range interactions were assumed to be present in the ferromagnetic metal being investigated. For instance in Fe, measurements of D, performed in a wide temperature range with triple axis spectrometry and small angle scattering[37] could be satisfactorily interpreted only with an itinerant electron model. Another support to this model in Fe, has been given by measurements of intensity of neutrons scattered by spin waves, as a function of energy[38]. It has been found that intensity falls off slowly with increasing energy up to 85 meV where the intensity drops by more than an order of magnitude (Figure 31).

It has been assumed that such a sudden decrease of the intensity is caused by the intersection of the spin wave spectrum with the continuous band of Stoner excitations. The position of intersection agrees fairly well with calculations of Stoner excitations as determined from the energy bands for iron.

Also in Ni similar results were obtained, not only as far as the behaviour of the dispersion curve is concerned, but also for the temperature dependence and the intensity decrease at high energy. The itinerant model must be also invoked for explaining the results in ferromagnetic alloys with the only remarkable exception of the Fe-Ni system which indicated fair agreement with a localized model. In neutron scattering experiments, antiferromagnetic systems have the advantage that purely magnetic peaks are present and therefore the difficulties connected with the scattering of nuclear origin are greatly reduced. This explains why the early studies of spin waves by inelastic neutron scattering were mostly performed in antiferromagnets. For example $RbMnF_3$ is a very simple antiferromagnet as the Mn ions are located on an undistorted simple cubic lattice, with spins parallel to a unique direction. The dispersion law of spin waves, determined either by triple axis and time of flight spectrometry[39] in the entire Brillouin[39] zone for the principal symmetry directions, has been found to be isotropic (Figure 32). The nearest neighbours exchange integral has been found to be - 0.29 MeV while second and third neighbours interactions are negligible. The temperature dependence has been also accurately studied. It indicates at constant $\underline{q}$ a decrease of the spin wave energy and a broadening of the width of the intensity peaks with increasing temperature, providing information on spin wave interactions and life-time.

Great attention has been paid to magnetite (Fe_3O_4) which can be considered the ferrimagnet more extensively investigated. In magnetite, which has a cubic spinel structure, eight of the tetrahedral sites (A sites) are occupied by Fe^{+3} and Fe^{+2} ions. Because of a rapid exchange of electrons all ions on B sites can be considered equal. Thus one expects three different exchange interactions J_{AA}, J_{BB} and J_{AB}.

Theoretical calculations indicated that the dispersion law consisted of one acoustic and five optic branches [23,40]. Their behaviour is of course determined by the values of the exchange interactions present in the system. These can be obtained experimentally if not only the acoustic branch is measured but the optic branch as well. Actually magnetite was the first system where optic modes of spin waves

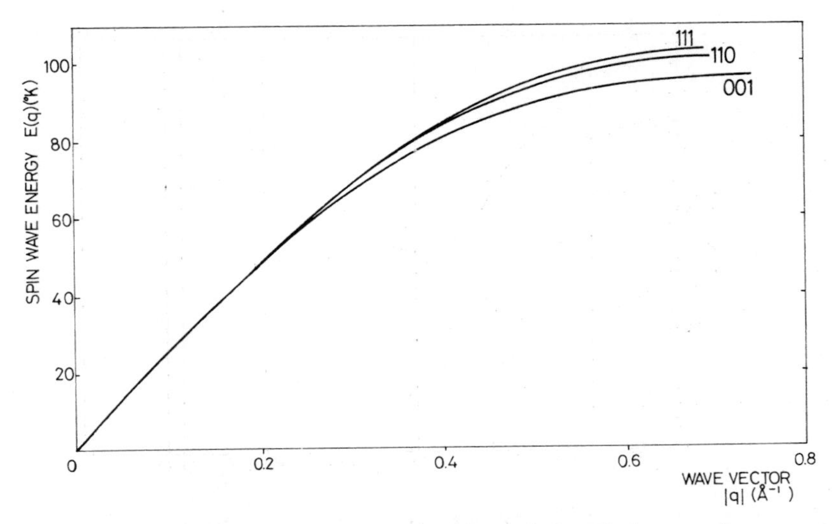

Figure 32 Dispersion curve of $RbMnF_3$ (from C.G. Windsor and R.W.H. Stevenson: (39a)) .
(Reproduced by permission of Academic Press Inc. (London) Limited)

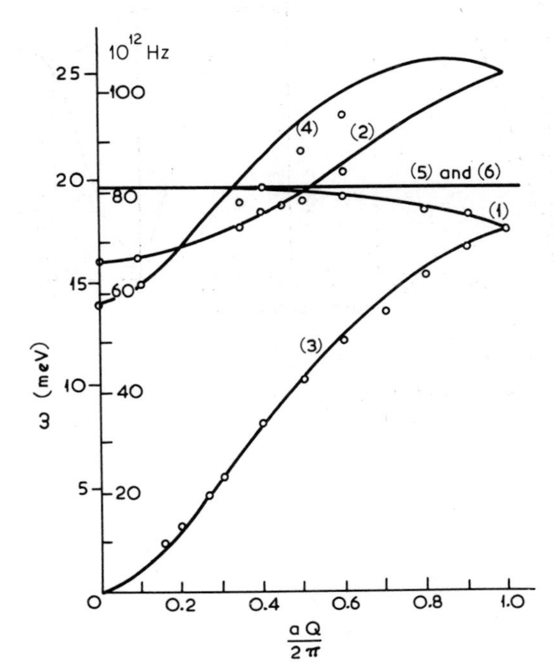

Figure 33 Dispersion curve of spin waves in magnetite (52).
(Reproduced by permission of Journal de Physique)

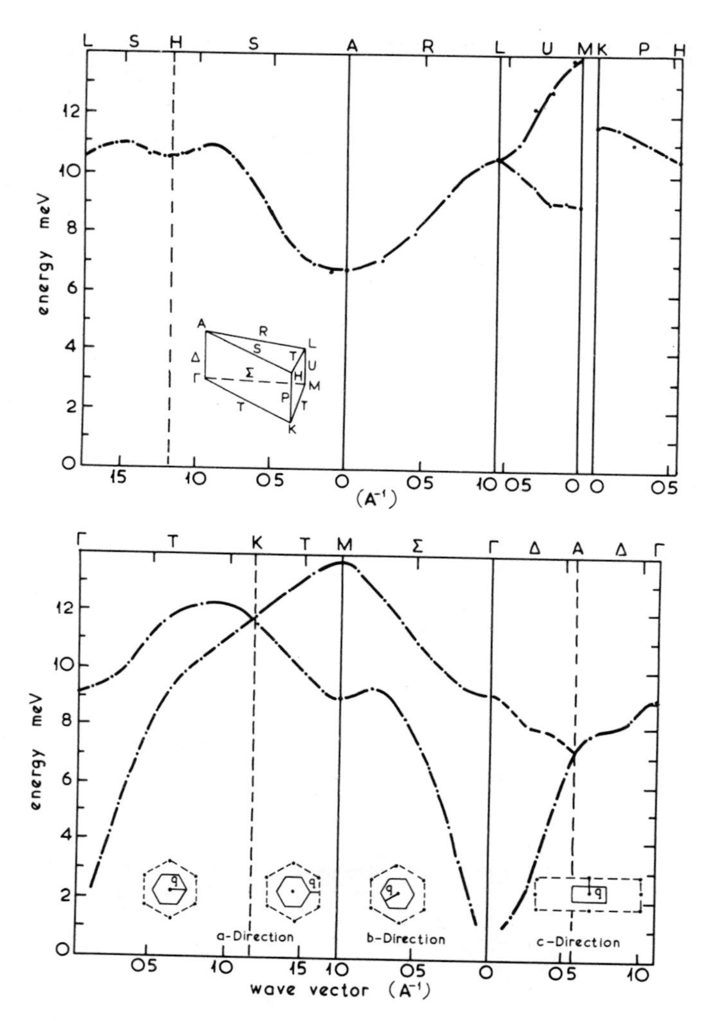

Figure 34 Magnon dispersion curves in Tb (from H.B. Møller et al (43b)).

were ever observed with a triple axis spectrometer [41]. More recently a very accurate investigation[42] with the same technique disclosed new optic branches, providing a full picture of the dispersion curve along the [001] direction (Figure 33). On the basis of these experimental results the J_{AA}, J_{BB} and J_{AB} exchange integrals could be evaluated. It turned out J_{AA} = -1.56 meV; J_{BB} = 0.26 meV and J_{AB} = -2.38 meV. It must be noticed that the B-B interactions is the only one ferromagnetic. The results agree with calculations based upon molecular field theory. Indications of pure influence of second neighbour interaction and deviations from a pure Heisenberg model were also found.

2) Metals and Compounds of Rare Earths

In rare earths the exchange interaction arises from an indirect interaction between highly localized 4f electrons via the conduction electrons. The markedly non spherical charge distribution of 4f electrons gives rise to an anisotropic crystal field and introduces also an anisotropy in the exchange interaction. Absorption cross sections of thermal neutrons are usually very high in rare earth, however in some cases, Tb, Ho and Er, the situation is more favourable also because of the presence of large magnetic moments which increase the magnetic contribution to the scattering. Moreover the magnon energies are low, so that it has been possible to measure the spin wave dispersion curves through the entire Brillouin zone. All the experiments on rare earths have been performed with triple-axis spectrometers.

In the ferromagnetic phase of Tb metal (at 90°K), the dispersion relation consists of an acoustic and an optic branch. As the 4f interaction is long range, the approximation (6") is not valid and the more general relation (5) must be used. The magnon dispersion relations have been measured along all the symmetry lines of the zone[43], the results are given in Figure 34. However, being the indirect exchange interaction in Tb long range, the measurements along symmetry lines are inadequate to magnon energies throughout the Brillouin zone and an analytical interpolation scheme had to be developed from which constant energy contours were obtained both for acoustic and optic magnons [44]. From these contours the magnon density of states may be constructed since the area between two contour lines of different energy is related to the number of states between the two energies. The results for Tb at 90°K are given in Figure 35. The magnetic contribution to the specific heat can then be calculated and a result of 7.3 ± 0.1 J/mole °K is obtained in fair agreement with the experimental value. The magnon energies in the c direction in Tb have been also studied as a function of temperature. They decrease with increasing temperature because of the magnon-magnon interaction.

A strong magnon-phonon interaction has been observed[45] in the alloy Tb-10% Ho in the ferromagnetic phase as shown in Figure 36. At the crossing of the magnon and the phonon dispersion curves is evident the existence of mixed magnon-phonon modes which could not be detected in pure Tb. In the Tb-10% Ho alloy the temperature dependence of the energy gap, i.e. the magnon energy at q = 0 was also studied. Since the energy gap depends only on the anisotropy constants, the results provide information on the anisotropy mechanism which is effective in the system (Figure 37). Also the field dependence of the spin wave energy spectrum at various $\underline{q}$, has been determined, which is related to the anisotropic coupling between the magnetic ions[46]. Among rare earth oxides EuO and EuS assume a particular interest, as they provide good examples of simple, isotropic Heisenberg ferromagnets. Unfortunately Europium has a very high absorption cross-section for neutrons and until lately neutron scattering measurements could not be performed. Only recently the use of samples prepared with the Eu^{153} isotope, made possible the study of spin waves in policrystalline material[47]. In general the scattering from a powder will show for a certain $\underline{q}$, a finite energy width because of the random orientation of the crystallites. The energy corresponding to the line peaks were fitted to the peak positions as they would be obtained by averaging equation (5) over all directions of $\underline{q}$ and using as parameters the exchange constants, limited to nearest and next nearest neighbours interactions J_1 and J_2. In this way it could be ascertained that in EuS $J_1 > 0$ and $J_2 < 0$ as obtained from specific heat and NMR data. In EuO however both the exchange constants are positive in contrast with specific heat and NMR experiments, but in agreement with theoretical calculations.

3) Two dimensional systems

Of particular interest is the information provided by the study of spin waves in bidimensional magnetic systems. In K_2NiF_4 the magnetic interactions exhibit a planar behaviour, as the spins are

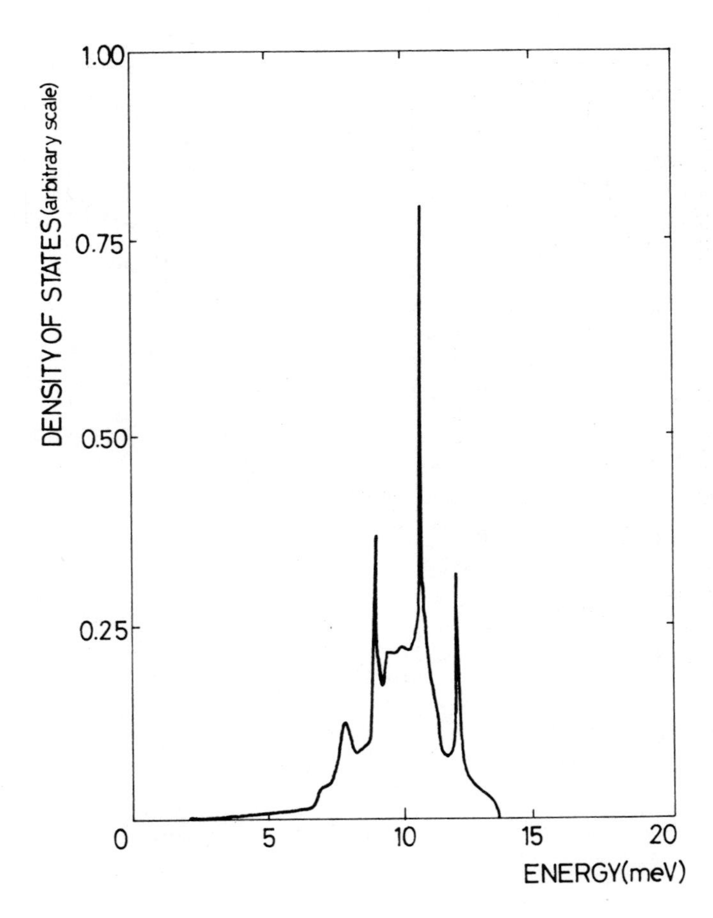

Figure 35 Magnon density of states in Tb: from (J.C.G. Houmann: (44)). (Reproduced by permission of International Atomic Agency)

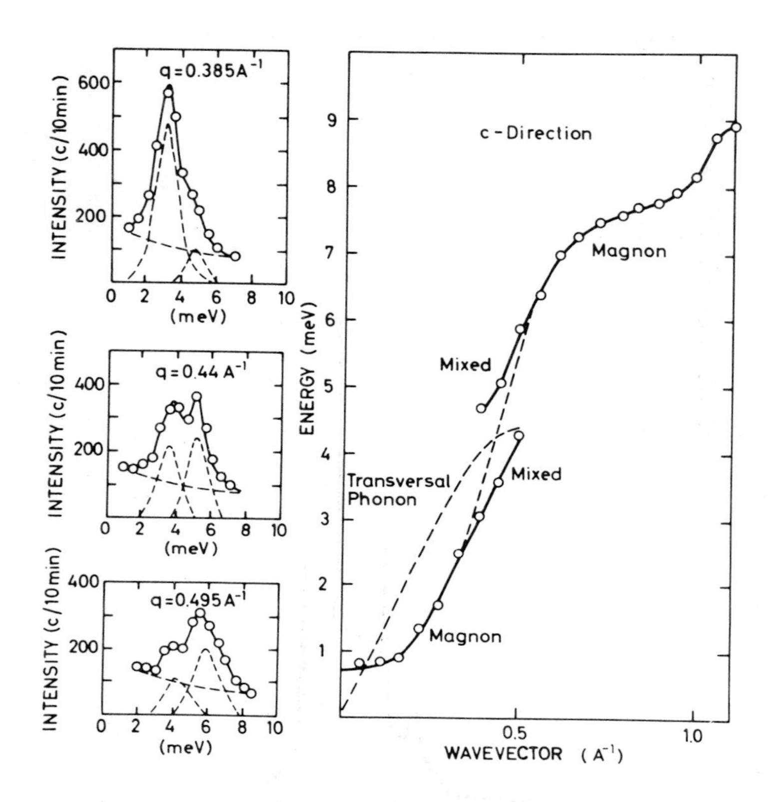

Figure 36 Experimental evidence of magnon-phonon interaction in the Tb-10% Ho alloy in the ferromagnetic phase (from H.B. Møller et al. (43b)).

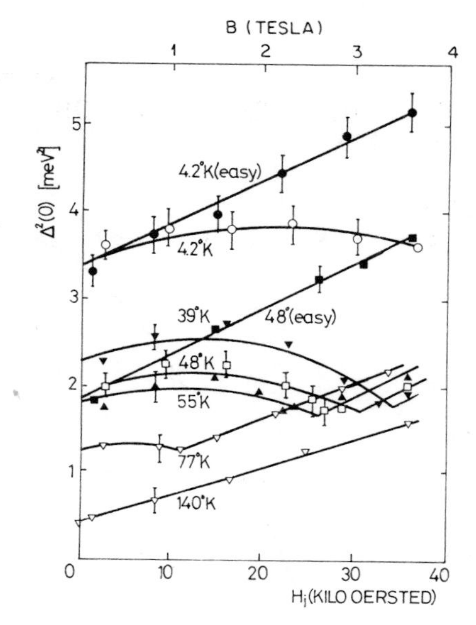

Figure 37 Temperature dependence of the magnon energy gap in Tb-10%Ho (from M. Nielsen et al. (53)).

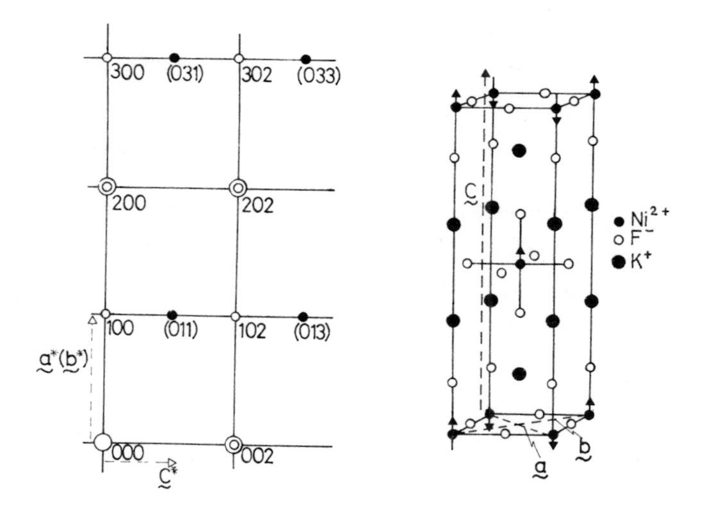

Figure 38 Crystal structure for K_2NiF_4(from J. Skalyo et al (49)).

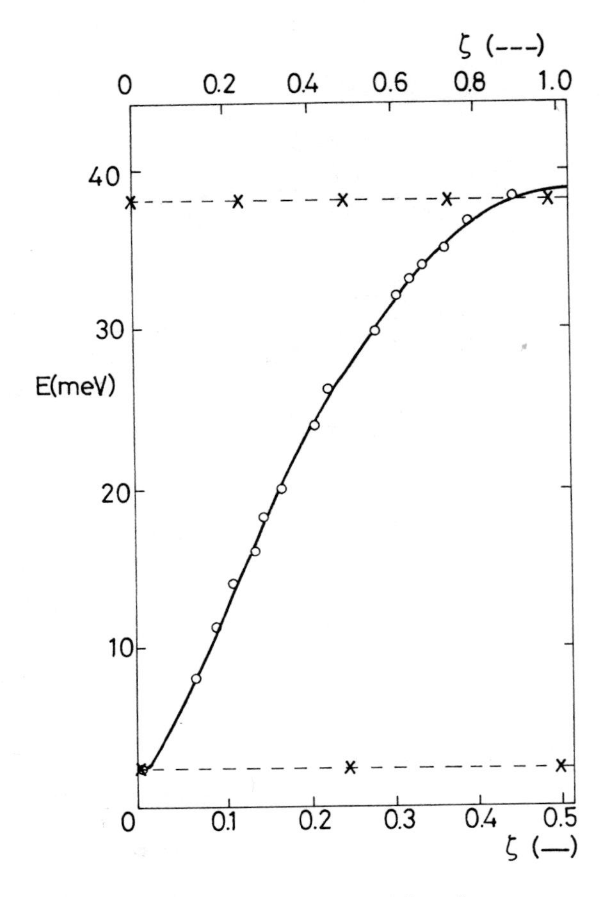

Figure 39 Magnon dispersion curves in the [010] zone for K_2NiF_4 (from J. Skalyo et al (49)). The solid line is for [ζ,0,0] while the dashed lines are for [0,0,ζ] at 2.4 meV and [0.45, 0,ζ] at 38.1 meV.

correlated within NiF_2 planes separated from one another along the c axis by two KF planes (Figure 38). An investigation of the magnons by neutron scattering can give quantitative information about the isotropy of the Hamiltonian and the relative strength of intra and interplanar interactions. In other words the bidimensional character of K_2NiF_4 would imply an anisotropic spin wave spectrum in different crystallographic directions. The magnon dispersion curve for [ζ00] and [00ζ] directions are given in Figure 39. The measurements in the c direction indicate the absence of measurable dispersion of the spin waves. On the basis of these results, the nearest neighbour interplanar exchange parameter could be estimated to be about 270 times smaller than the one in the plane , giving a strong support to the two dimensional picture of K_2NiF_4. In the temperature dependence of the [100] magnon peak, no measurable change of the width up to $T = 1.1\ T_N$ was observed. This result is consistent with quasi elastic scattering of neutrons indicating the existence of long range two-dimensional spin correlations over a wide temperature interval above the phase transition.

Conclusion

We wish to point out that in this field, as in several others in physics, experimental methods which "a priori" can be recommended as the best ones for solving these problems, just do not exist. Furthermore the situation is complicated, by the fact that the probes which are being used, i.e. neutrons, can be generated by different types of sources, not only as far as the intensity is concerned but also regarding the temporal dependence of the flux (continuous or pulsed sources). Also the geometry of the beam holes or the existence of particular facilities like hot and cold neutron sources, neutron guiding tubes etc., play a major role.

Some general rules can be given in some cases: polarized neutrons are very useful when dealing with ferromagnets, time of flight techniques have to be used for pulsed sources; but the experimentalist can decide only after comparing the kind of results which can be obtained by the different techniques, including the limits of these results (energy and momentum range, resolution, statistical errors), with the information he is looking for. The same consideration apply of course to the sometimes overlooked problem of giving the proper credit to the experimental results obtained with different methods.

For this reason a good familiarity with the different experimental methods is necessary and in a rather detailed fashion. This review aimed only to provide a first approach to the problem and, for those who plan experiments in this field, a stimulus for gaining a deeper insight, through the original literature which has been mentioned.

References

1. (a) R.M. Moon, Phys. Rev. 136, (1964)
   (b) R.M. Moon, Int. J. Magnetism 1, 219 (1971)
   (c) C.G. Shull and H.A. Mook, Phys. Rev. Letters 16, 184 (1966)
2. A. Paoletti and F.P. Ricci, Nucl. Inst. Methods, 26, 125 (1964)
3. See for instance H. Maier Leibnitz "Neutron Inelastic Scattering" IAEA, Vienna 1972 p 681.
4. T.T. De Marco, Phil. Mag. 16, 1303 (1967)
5. G. Caglioti, A. Paoletti, F.P. Ricci, Nucl. Inst. Methods, 9, 195 (1960).
6. R.M. Moon, IV Sagamore Conference, Minsk 1973, unpublished.
7. R. Nathans, A.J. Freeman and R.E. Watson, J. Appl. Phys., 34, 1182 (1963).
8. (a) R. Nathans et A. Paoletti, Phys. Rev. Letters, 2, 254 (1959)
   (b) R. Nathans, C.G. Shull, G. Shirane and A. Andresen J. Phys. Chem. Solids 10, 138 (1959)
9. C.G. Shull and Y. Yamada, J. Phys. Soc. Japan suppl. BIII p. 1 (1962)
10. For more extensive reviews of experimental and theoretical results on magnetization densities see for instance
    (a) A. Paoletti, Riv. Nuovo Cimento 2, 451 (1970) and
    (b) R.M. Moon, Trans. Am. Cryst. Ass. 8, 59 (1972).
11. C.G. Shull and M.K. Wilkinson, Phys. Rev., 97, 304 (1953)
12. (a) B. Antonini, F. Menzinger, A. Paoletti and F. Sacchetti, Int. J. Magnetism, 183 (1971).
    (b) F. Leoni, F. Menzinger and F. Sacchetti, Solid State Communications, 13, 775 (1973).
13. (a) J. Kanamori,J. de Physique Suppl. 35, C4-131 (1974)
    (b) Y. Nakai and J. Kunitomi, J. Phys. Soc. Japan 39, 1257 (1975)
    (c) F. Sachetti, P. De Gaspenis, F. Menzinger, Phys.Stat.Solidi B 76, 309 (1976).
14. B. Antonini, F. Lucari, F. Menzinger, A. Paoletti, Phys.Rev.,187 611 (1969)
15. F. Menzinger, F. Sacchetti, M. Romanazzo, Phys.Rev., 5, 3778 (1972)
16. M. Bonnet, A. Delapalme, F. Tcheon and H. Fuess, IV Sagamore Conference, Minsk 1973, unpublished.
17. R.M. Moon, W.C. Koehler, J.W. Cable and H.R. Child,

Phys. Rev., 5, 997 (1972).

18. R.M. Moon and W.C. Koehler, Proc. Int. Conference of Magnetism I.C.M. 73 (Publishing House Nauka Moscow, 1974)

19. (a) D.E. Cox, E.J. Sammelsen and K.M. Beckurts, Phys. Rev., 7, 3102 (1973)
 (b) F. Leoni and F. Sacchetti Phys. Rev., 7, 3112 (1973).

20. For a review on spin waves in the Heisenberg model see for instance
 (a) J. Van Kranendonk and J.H. Van Vleck, Rev. Mod. Phys., 1, 30 (1958)
 (b) F. Keffer, Handb. Phys., 18, 2 (1967)
 Spin waves from the itinerant electrons point of view are treated by
 (c) C. Herring in Magnetism (Ed. G.T. Rado and H. Suhl) Academic Press, New York, 1966.

21. (a) F. Bloch, Z. Physik, 61, 206 (1930),
 (b) J.C. Slater, Phys. Rev., 35, 509 (1930).

22. (a) B. Buras,"Theory of Condensed Matter"IAEA Vienna 1968 p. 473
 (b) ibid p. 478.

23. See for instance T.A. Kaplan, Phys. Rev., 109, 782 (1958)

24. B.N. Brockhouse, "Inelastic Scattering of Neutrons in Solids and Liquids" IAEA Vienna 1961 p. 113.

25. R.N. Sinclair and B.N. Brockhouse, Phys. Rev., 120, 1638 - 40 (1960).

26. B.N. Brockhouse, C.A. de Witt, E.D. Hallman and J.M. Rowe "Neutron Inelastic Scattering" IAEA Vienna 1968 Vol. II p. 259

27. (a) "Neutron Inelastic Scattering" IAEA Vienna 1968, Session on Experimental Methods, Vol. II p. 251,
 (b) "Neutron Inelastic Scattering " IAEA Vienna 1972, Session on Instrumentation, p. 679.
 For the resolution of a triple axis spectrometer see also
 (c) M.J. Cooper and R. Nathans, Acta Crystall., 23, 357 (1967)

28. O.K. Harling, " Neutron Inelastic Scattering" IAEA Vienna p. 271

29. (a) L. Pal, N. Kroo, P. Pellionisz, F. Szlavik and I. Vizi "Neutron Inelastic Scattering" IAEA Vienna 1968 p. 407
 (b) N. Kroo, P. Pellionisz, I. Vizi, G. Zsigmond, G. Zhukov "Neutron Inelastic Scattering" IAEA Vienna 1972 p 763

30. R. Amadori and F. Hossfeld, "Neutron Inelastic Scattering" IAEA Vienna 1972 p. 747.

31. O. Steinsvoll and A. Virjo, "Neutron Inelastic Scattering" IAEA Vienna 1968 p. 395.

32. (a) R.J. Elliott and R.D. Lowde , Proc. Poy. Soc., A 230, 46 (1955) and

(b) H.A. Gein, H.A. Alperin, O. Steinsvoll, R. Nathans and G. Shirane, Phys. Rev., 154, 508 (1967)

33. (a) Ann. Tech., Report of Institut for Atomenergy, Kjeller, Norway, N. 3 (1963) upublished,

(b) A. Wanic, Nukleonika, 9, 839 (1964)

34. A.D.B. Woods et al "Neutron Inelastic Scattering" IAEA Vienna, 1968 p. 281

35. B. Antonini, F. Menzinger and R. Medina, Nucl. Instr. Methods, 87, 125 (1970).

36. B. Antonini, F. Menzinger, Solid State Comm. 9, 417 (1971)

37. M.W. Stringfellow, J. Phys. C., 1, 950 (1968).

38. H.A. Mook and R.M. Nicklow, Journ. de Physique, 32, Suppl. C 1 -1177 (1971).

39. (a) C.G. Windsor and R.W.H. Stevenson, Proc. Phys. Soc., 87, 501 (1966)

(b) D.H. Saunderson C, C.G. Windsor, G.A. Briggs, "Neutron Inelastic Scattering" IAEA Vienna, 1972 p. 639

40. M.L. Glasser and F.J. Milford, Phys. Rev., 130, 1783 (1963). References to previous work are given.

41. H. Watanabe and B.N. Brockhouse, Physics Letters, 1, 189 (1962)

42. Group de Diffusion Inelastique des Neutrons, J. de Physique, 32, Suppl. C 1, 1182 (1971).

43. (a) H.B. Møller and J.C.G. Houmann, Phys. Rev. Letters 16, 737 (1966)

(b) H.B. Møller, J.C.G. Houmann and A.R. Mackintosh, J. Appl. Phys., 39, 807 (1968).

44. J.C.G. Houmann, "Neutron Inelastic Scattering" IAEA Vienna 1968 Vol. II p. 29.

45. H.B. Møller, J.C.G. Houmann and A.R. Mackintosh, Phys. Rev. Letters, 19, 312 (1967)

46. H.B. Møller, J.C.G. Houmann, J. Jensen and A.R. Mackintosh

"Neutron Inelastic scattering" IAEA Vienna, 1972 p. 603.

47. L. Passel, O.W. Dietrich and J. Als-Nielsen, AIP Conf. Proc. 5 (Magnetism and Magnetic Materials 1971) Part II p 1251 (1972)

48. S.J. Pickart, A.H. Alperin, G. Shirane and R. Nathans, Phys. Rev. Letters, 12, 444 (1964).

49. J. Skaylo, G. Shirane, R.J. Birgeneau and H.J. Guggenheim, Phys. Rev., Letter 23, 1394 - 1397 (1969)

50. H.B. Møller in "Neutron Inelastic Scattering" IAEA Vienna 1968 Vol. II p. 5

51. G. Shirane, V.J. Minkiewicz and R. Nathans, J. Appl. Phys., 39, 383-390 (1968).

52. Group de Diffusion Inelastique des Neutrons, J. de Physique 32 Suppl. Cl 1182 - 3 (1971).

53. M. Nielsen, H.B. Møller, P.A. Lindgard and A.R. Mackintosh, Phys. Rev. 12, 332 (1975).

Bibliography

B1 G.E. Bacon "Neutron Diffraction" 3rd Edition
Clarendon Press, Oxford 1975

B2 P.A. Egelstaff, (Ed) "Thermal Neutron Scattering"
Academic Press, London 1965

B3 C. Kittel "Introduction to Solid State Physics"
3rd Edition, John Wiley 1966

B4 W. Marshall and S.W. Lovesey "Theory of Thermal Neutron
Scattering" Clarendon Press, Oxford 1971

B5 D.C. Mattis "The Theory of Magnetism" New York 1965

B6 R.J. Weiss "X ray Determination of Electron Distributions"
North Holland - Wiley Interscience, New York 1966.

Magnetic Anisotropy

R.F. PEARSON
Philips Laboratories, Redhill

1. Introduction

The magnetic anisotropy of ferromagnetic and ferrimagnetic crystals is now well known. Experimental measurements have shown that there are certain directions along which single crystals are more easily magnetized and in general they coincide with the principal crystal symmetry directions. For example in pure iron these are <100> directions, and for magnetite (Fe_3O_4) these are <111> directions at room temperature.

The free energy of the crystal thus contains a term, which is dependent on the magnetization direction relative to the crystallographic axis, and is known as the magnetocrystalline energy. The magnitude of the magnetocrystalline energy reflects the anisotropy of the crystal and is expressed in terms of parameters usually referred to as anisotropy constants.

In this Chapter we shall show how it is possible to derive the intrinsic anisotropy constants (as defined in the expression for the magnetocrystalline energy) from experimental measurements of the macroscopic magnetic properties of single crystals.

To illustrate this point, let us consider the case of a sample of magnetization M and magnetocrystalline anisotropy energy E_a which is placed in a magnetizing field B sufficiently large to magnetize it to saturation and remove all domain walls. The relative equilibrium orientations of $\underline{M}$ and $\underline{B}$ are shown in Figure 1.

$\underline{M}$ and $\underline{B}$ are assumed to make angle of θ and ϕ respectively with a coplanar direction $\underline{R}$ in the crystal. (Note that for most crystals there usually exist planes with high symmetry for which this condition will be satisfied). Then the total free energy per unit volume of the crystal (neglecting terms involving the demagnetization energy which depends on the shape of the sample) is given by

$$E = E_a - \underline{M} \cdot \underline{B} \cos (\phi - \theta) \tag{1}$$

For equilibrium $$\frac{\partial E}{\partial \theta} = 0 = - \underline{M} \cdot \underline{B} \sin (\phi - \theta) + \frac{\partial E_a}{\partial \theta} \tag{2}$$

Torque per unit volume $$= \underline{M} . \underline{B} \sin (\phi - \theta) = + \frac{\partial E_a}{\partial \theta} \tag{3}$$

Note that the above expression gives the torque per unit volume of the crystal due to the magnetic field.* Thus we can relate the torque and the normal component of the magnetization to the variation of anisotropy energy E_a in the crystal plane under consideration.

Another method of describing the anisotropy is in terms of the effective stiffness binding the magnetization to particular crystallographic directions. The stiffness C is given by

$$\text{Stiffness } C = \left(\frac{\partial^2 E_a}{\partial \theta^2}\right)_{\theta=0} \tag{4}$$

The stiffness can be anisotropic and will be different for deflection of M in different directions.

We can also express this stiffness as an anisotropy field B_A where B_A is given

$$B_A = \frac{C}{M} = \frac{1}{M}\left(\frac{\partial^2 E_a}{\partial \theta^2}\right)_{\theta=0} \tag{5}$$

and this is a measure of the magnetizing field required to rotate the magnetization vector out of an easy direction.

In ferromagnetic resonance experiments where the magnetization vector precesses around a particular applied field direction, the effective field due to anisotropy energy is given by

$$B_{eff} = \frac{1}{M\sin\theta}\left[\left(\frac{\partial^2 E}{\partial \theta^2}\right)\left(\frac{\partial^2 E}{\partial \phi^2}\right) - \left(\frac{\partial^2 E}{\partial \theta\,\partial \phi}\right)^2\right]^{\frac{1}{2}} \tag{6}$$

where θ and ϕ are the polar and azimuthal coordinates of the magnetization.

The resonance frequency ω is then given by $\omega = \gamma B_{eff}$. where γ is the gyromagnetic ratio.

* the torque per unit volume due to the crystal anisotropy is then given by $|L| = -\partial E_a/\partial\theta$.

2. Magnetic properties of a uniaxial crystal

If we take the very simple case of a crystal with only uniaxial anisotropy we can calculate the effect of magnetic anisotropy on the torque curve, magnetization curve and microwave resonance field respectively. This will assume that the anisotropy is independent of frequency of motion of the magnetization. (The case of a frequency dependent anisotropy is treated in Section 9). For uniaxial anisotropy, the anisotropy energy depends only on the angle between the magnetization and the reference or uniaxial preferred direction of magnetization and can be expressed in simplest form as:

$$E_a = K \sin^2\theta \tag{7}$$

where θ is the angle between $\underline{M}$ and the reference direction $\underline{R}$ as shown in Figure 1, K is a constant, referred to usually as the anisotropy constant. If K is positive, this anisotropy energy has a minimum at $\theta = 0$, that is when $\underline{M}$ is pointing along the direction $\underline{R}$,which is thus the easy direction of magnetization.

If we consider the torque variation in a plane through the crystal containing the direction $\underline{R}$, we get the torque due to anisotropy as follows

$$L = -\left(\frac{\partial E_a}{\partial \theta}\right) = -2\ K \sin\theta \cos\theta = -K \sin 2\theta \tag{8}$$

Now if we plot the torque per unit volume against magnetization direction as shown in Figure 2, we obtain a sin 2θ curve with peak height equal to K the uniaxial anisotropy constant. At the points $\theta = 0$ and $\theta = 180$, where $\underline{M}$ points along the direction $\underline{R}$, the slope of the torque curve is given by -K, which for positive values of K is negative. Note if $\underline{R}$ is a hard direction, the slope would be reversed in sign (i.e. positive), and indicating a negative value of K.

Now let us also consider the magnetization curve for a simple uniaxial crystal.

For B applied along the easy direction, the sample should ideally saturate in fields above zero.

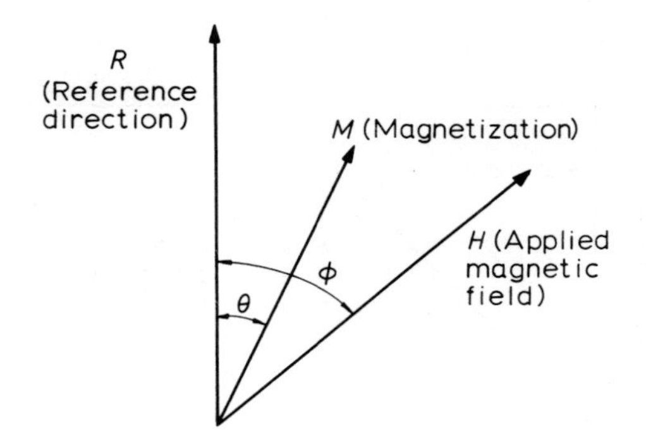

Figure 1 Orientation of crystal magnetization and applied field directions.

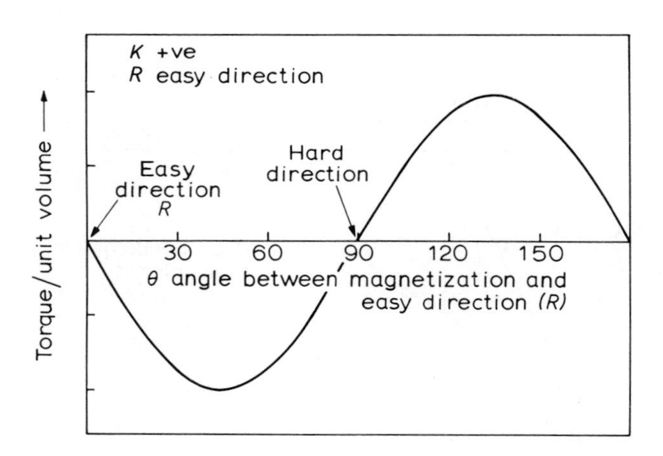

Figure 2 Torque curve for a crystal with uniaxial anisotropy in a plane containing the uniaxial direction.

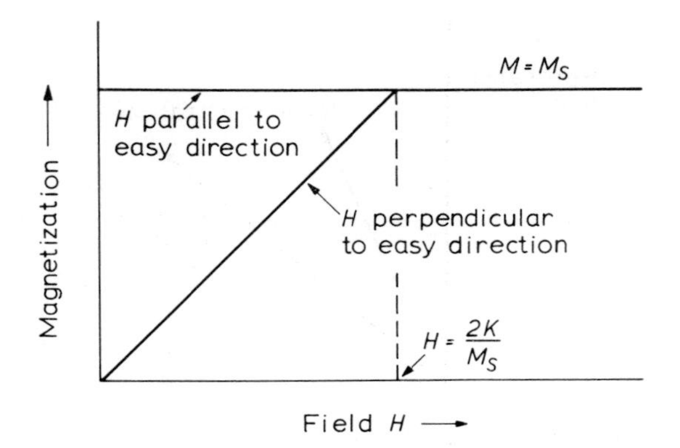

Figure 3 Magnetization curves of a uniaxial crystal, parallel and perpendicular to the easy direction.

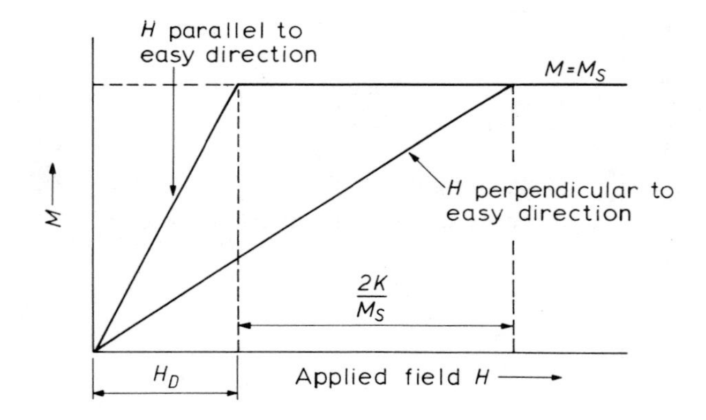

Figure 4 Effect of demagnetizing field on the magnetization curve of a uniaxial crystal.

Thus

$$M = M_s \text{ in all fields greater than } B = 0$$

Note that in practice, magnetization curves of finite initial slope are obtained due to the demagnetizing factor of the crystal. For example when applied perpendicular to R we get regions of magnetization (domains) parallel to the R_{axis} which rotate by equal amounts towards the applied field direction. Thus if we rewrite, equation 8 and 3 in the form

$$K \sin 2\theta = M_s B \sin(\phi - \theta) \qquad (9)$$

for $\phi = 90$

$$2K \sin\theta \cos\theta = M_s B \cos\theta \qquad 2K \sin\theta = M_s B \qquad (10)$$

The magnetization along the field direction B is then given by

$$M = M_s \cos(\phi - \theta) \qquad M_s \sin\theta = \frac{M_s^2 B}{2K} \qquad (11)$$

Note that for this direction

$$M = M_s \text{ when } B = \frac{2K}{M_s}$$

and this is referred to as the effective anisotropy field B_A.

These curves are shown in Figure 3.

However in practice, one has to take into account the demagnetizing factor of the crystal. The effect of this is to introduce an effective demagnetizing field $B_D = \mu_o D M_s$. The magnetization curves are then shown in Figure 4.

For fields applied at intermediate angles to the easy direction, the magnetization curve (for the single domain state) is given by

$$M = M_s \cos(\phi - \theta) \qquad (12)$$

where

$$K \sin 2\theta = M_s B \sin(\phi - \theta)$$

Therefore putting $\theta = \theta_1$ that is the direction of interest we obtain

$$\frac{K \sin 2\theta}{M_s B} = \left[1 - \left(\frac{M}{M_s}\right)^2\right]^{\frac{1}{2}} \qquad (13)$$

Note that for these cases ultimate saturation, i.e. $M = M_s$ is <u>never</u> reached for finite values of B.

We can also calculate the effect of anisotropy in the field required for ferromagnetic resonance.

Using the expression given in equation 6 we can derive the field for resonance in the easy and hard directions for a uniaxial crystal.

$$B_r \text{ - along the easy direction} = \left(\frac{2K}{M_s} + B\right) \qquad (14)$$

B_r - perpendicular to the easy direction (15)

For $B < 2K/M$ $\qquad B_r{}^2 = \left(\frac{2K}{M_s}\right)^2 - B^2$

For $B > 2K/M$ $\qquad B_r{}^2 = B\left(B - \frac{2K}{M_s}\right)$

The resonance frequency is plotted as a function of applied field for both the easy and hard directions in Figure 5.

In these series of simple examples we have seen how expressions for the torque, magnetization and field required for microwave resonance can be related to the anisotropy constants of the material for the simple case of a uniaxial crystal. In the following sections this analysis is extended to more complicated anisotropy energies that can be encountered in practice and also describes in some detail experimental techniques for obtaining the data, and the methods used to analyse the experimental results.

3. Anisotropy energy of cubic and other systems

If we consider that part of the free energy of the crystal that is associated with the magnetocrystalline anisotropy and depends on the direction of the magnetization , then in the case of cubic crystals this energy may be expressed as:

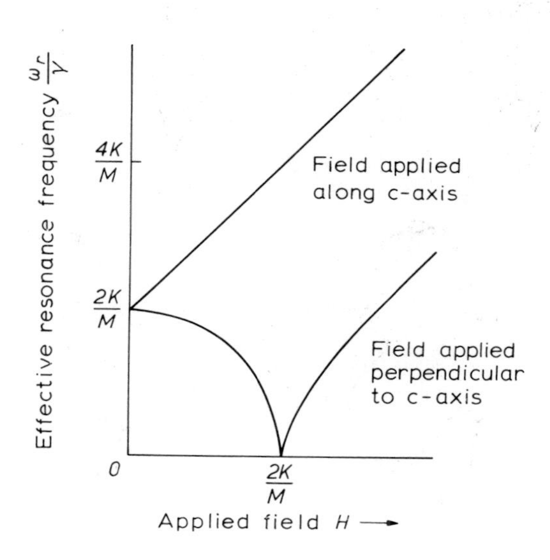

Figure 5 Ferromagnetic resonance frequency as a function of applied field for a spherical crystal with uniaxial anisotropy along the easy and hard directions.

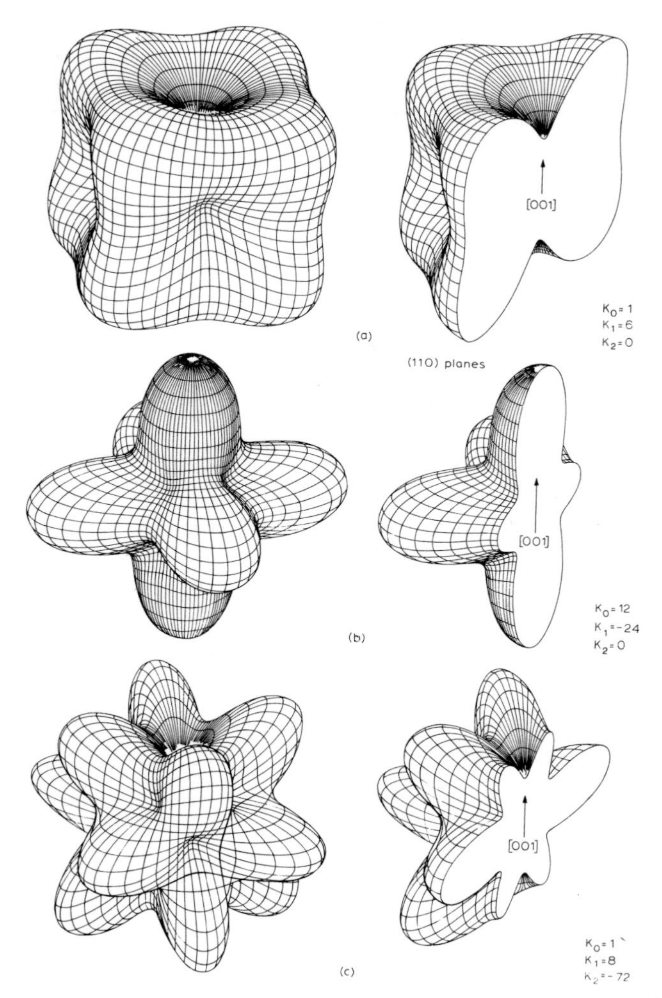

Figure 6 Perspective drawings of the magnetocrystalline energy surface for cubic crystals.

(a) K_1 + ve, $K_2 = 0$

(b) K_1 - ve, $K_2 = 0$

(c) $K_2 = -9K_1 = -ve$

$$E_a = K_0 + K_1(\alpha_1{}^2\alpha_2{}^2 + \alpha_2{}^2\alpha_3{}^2 + \alpha_3{}^2\alpha_1{}^2) + K_2\alpha_1{}^2\alpha_2{}^2\alpha_3{}^2 \qquad (16)$$

where α_1, α_2, α_3, are the direction cosines of the magnetization with respect to the cubic axes and K_0, K_1, K_2 are the magnetocrystalline anisotropy constants , which are generally used for cubic crystals.

This expression satisfies cubic symmetry as it only contains even powers of the direction cosines of the magnetization and is unaltered by either interchanging the direction cosines or by changing their signs.

Alternative expressions using symmetric combinations of spherical harmonics have also been suggested.[1,2] However although these may be useful in special cases where higher order coefficients are required, most experimenters still prefer the method we have adopted here.

If we now plot this crystal energy against direction of magnetization it shows maxima and minima along certain crystal directions. These crystal directions depend on the values of K_1 and K_2 the anisotropy constants.

For positive values of K_1 and values of K_2 much less than K_1 the energy minima lie along the cubic edges <100> for which $\alpha_1 = \alpha_2 = 0$ $\alpha_3 = 1$. For negative values of K_1 (K_2 small) the energy minima lie along the <111> directions, e.g. cube diagonals for which $\alpha_1 = \alpha_2 = \alpha_3 = 1/\sqrt{3}$.

The variation in magnetocrystalline energy for all directions can be illustrated by plotting the energy as a three dimensional surface in which the distance from the surface to the origin represents the energy for the magnetization pointing along that direction. Perspective drawings for the surface with cubic anisotropy and particular values of K_1, K_2 as given are shown in Figure 6. Sections of the anisotropy surface give the anisotropy energy in particular crystal planes.

The energy curves in Figure 6 were drawn using computed values of a simple projection of the three dimensional energy surface, and plotting only those points which would be visible from the point of projection.

For appreciable values of K_2, that is when higher order terms are taken into account, certain other directions than <100> or <111> may become the easy direction of magnetization. This is shown graphically in Figure 7, from which the easy direction for all possible values of K_1 and K_2 may be decided. The results are summarised in Table 1.

TABLE 1

Directions of easiest, intermediate and hardest magnetization in cubic crystals

K_1	+	+	+	−	−	−
K_2	$-\infty$ to $-9K_1/4$	$-9K_1/4$ to $-9K_1$	$-9K_1$ to $-\infty$	$-\infty$ to $9\|K_1\|/4$	$9\|K_1\|/4$ to $9\|K_1\|$	$9\|K_1\|$ to $+\infty$
Easiest	<100>	<100>	<111>	<111>	<110>	<110>
Intermediate	<110>	<111>	<100>	<110>	<111>	<100>
Hardest	<111>	<110>	<110>	<100>	<100>	<111>

Although the simple expression given in equation 16 is usually found adequate to express the experimentally observed anisotropy of cubic crystals (e.g. iron, nickel, spinel ferrites, garnets) it is of interest to outline a more general way by which expressions of a similar form suitable for crystals with different symmetry may be derived (e.g. hexagonal Co, hexagonal ferrites, α-Fe_2O_3 etc.)

If we consider the case of a perfectly rigid crystal the crystal anisotropy energy E_a can be expressed in the following way:

$$E_a = b_i\alpha_i + b_{ij}\alpha_i\alpha_j + b_{ijk}\alpha_i\alpha_j\alpha_k + b_{ijk\ell}\alpha_i\alpha_j\alpha_k\alpha_\ell + \text{higher terms} \quad (17)$$

In this equation tensor notation is used and the coefficients b_i, b_{ij}, $b_{ijk\ell}$ are components of first, second and fourth rank tensors and are physical constants of the crystal. The direction cosines of the magnetization vector with respect to the crystal axes are given by $\alpha_{i,j,k,\ell}$. Now if we impose the requirements of the symmetry of

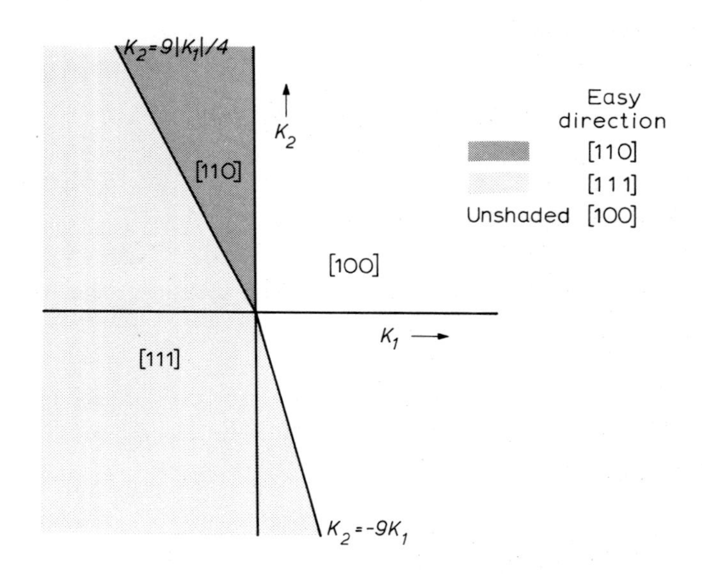

Figure 7 Directions of easy magnetization for cubic crystals as a function of the anisotropy constants, K_1, K_2

the crystal on the anisotropy energy expression in equation 16 we see that the components $b_{ijk\ell}$ must conform to the crystal symmetry. These components must be invariant when subjected to the operations which characterise the symmetry properties of the crystal under investigation . Thus each of the constants must obey transformations of the form shown below

$$b_{ijk\ell} = P_{ip}P_{jq}P_{kr}P_{\ell s}b_{pqrs} \tag{18}$$

where the quantities P_{ip} etc. are the symmetry operators for the particular type of crystal we are considering.

<u>Example</u>

For cubic crystals of class O^h, m^{3m}, the symmetry operators are

$$\begin{bmatrix} 1 & 0 & 0 \\ 0 & 0 & 1 \\ 0 & -1 & 0 \end{bmatrix} \quad \begin{bmatrix} 0 & 0 & 1 \\ 1 & 0 & 0 \\ 0 & 1 & 0 \end{bmatrix} \quad \begin{bmatrix} -1 & 0 & 0 \\ 0 & -1 & 0 \\ 0 & 0 & -1 \end{bmatrix} \tag{19}$$

If we now use these operators in equation 3 we find that all terms containing an odd number of subscripts vanish and we neglect the constant term we obtain the familiar expression

$$E_a = K_1(\alpha_1{}^2\alpha_2{}^2 + \alpha_2{}^2\alpha_3{}^2 + \alpha_3{}^2\alpha_1{}^2) \tag{20}$$

where K_1 depends only on the value of the b coefficients (Birss[1] p.155)

If higher order terms are required the series given by equation 17 must be extended.

The sixth order term is given by

$$b_{ijk\ell mn}\alpha_i\alpha_j\alpha_k\alpha_\ell\alpha_m\alpha_n$$

which results in the following expression

$$E_a = K_1(\alpha_1{}^2\alpha_2{}^2 + \alpha_2{}^2\alpha_3{}^2 + \alpha_3{}^2\alpha_1{}^2) + K_2\alpha_1{}^2\alpha_2{}^2\alpha_3{}^2 \tag{21}$$

An extension to even higher terms would give the next term of the form

$$K_3\ \alpha_1{}^2\alpha_2{}^2 + \alpha_2{}^2\alpha_3{}^2 + \alpha_3{}^2\alpha_1{}^2)\ ^2 \qquad (22)$$

Now let us consider the case of hexagonal crystals of class $D_6{}^h$, 6/mmm.

The symmetry operators for this class are

$$\begin{bmatrix} 1 & 0 & 0 \\ 0 & 1/2 & \sqrt{3}/2 \\ 0 & \sqrt{3}/2 & 1/2 \end{bmatrix} \cdot \quad \begin{bmatrix} -1 & 0 & 0 \\ 0 & 1 & 0 \\ 0 & 0 & 1 \end{bmatrix} \quad \begin{bmatrix} 1 & 0 & 0 \\ 0 & -1 & 0 \\ 0 & 0 & 1 \end{bmatrix} \qquad (23)$$

If we use the same approach as shown for cubic crystals above it can be shown (explained in detail Birss[1] p. 158) that the anisotropy energy can be represented by

$$E_a = K_1(\alpha_1{}^2 + \alpha_2{}^2) + K_2(\alpha_1{}^2 + \alpha_2{}^2)\ ^2$$

which reduces to

$$E_a = K_1 \sin^2\theta + K_2 \sin^4\theta \qquad (24)$$

where θ is angle between the c axis and the magnetization. We notice that this expression is dependent to fourth order only on θ and is therefore isotropic in the basal plane or hexagonal plane. If we wish to take into account the anisotropy in this plane we must go to higher order terms. It is convenient to change to polar coordinates θ, ϕ, where θ is the angle between the magnetization direction and the c-axis and ϕ is the angle between the projection of M in the basal plane and the a-axis in that plane. We then get the following expression

$$E_a = K_1 \sin^2\theta + K_2 \sin^4\theta + K_3 \sin^6\theta + K_4 \sin^6\theta \cos 6\phi \qquad (25)$$

If K_1 is large compared with the other constants, the preferred directions of magnetization are given by

K_1 is positive - c-axis is easy direction.

TABLE 2

Easy directions for hexagonal crystals

K_1	+	+	−	−
K_2	$+\infty$ to $-K_1$	$-K_1$ to $-\infty$	$-\infty$ to $-\frac{1}{2}K_1$	$-\frac{1}{2}K_1$ to $+\infty$
Easy direction	$\theta = 0$ (c-axis)	$\theta = 90$ (basal plane)		$\theta = \sin^{-1}(-K_1/2K_2)^{\frac{1}{2}}$ (easy cone)

The anisotropy surface is shown in Figure 8(a).

K_1 is negative - basal plane is easy

c-axis is hard direction

The anisotropy surface is shown in Figure 8(b).

For particular values where K_2 is of the same magnitude as K_1 it is possible to have an easy cone of directions which are all easy. The corresponding energy surface is shown in Figure 8(c). If K_4 is different from zero, six preferential directions 60° apart are found in the basal plane.

These results are conveniently summarised in Table 2.

Using the fairly general method outlined in this section, similar results can be derived for other crystal systems and the complete results for cubic, hexagonal, trigonal, orthorhombic and tetragonal systems are tabulated in Table 3.

These energy expressions have defined in a formal manner the anisotropy of the crystal in terms of various constants or coefficients. The task of the experimentalist is to determine the values of these constants and in the following sections the methods of deducing these constants from experimental parameters such as magnetization curves, torque curves, ferromagnetic resonance measurements will be described. Obviously the properties of the material under investigation and indeed the size and shape of the single crystal sample that can be prepared will determine the optimum method for determining the anisotropy constants.

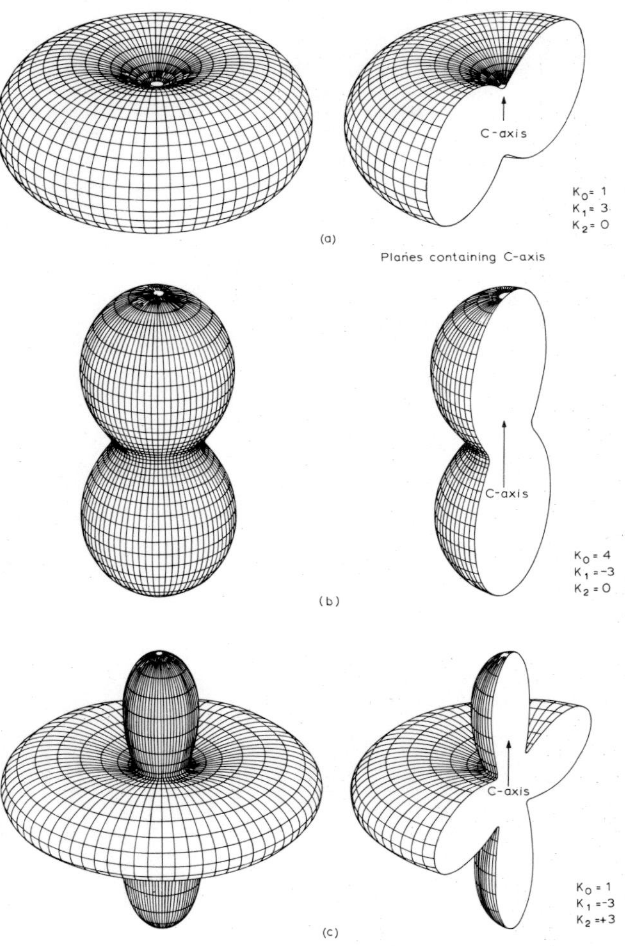

Figure 8 Perspective drawings of anisotropy energy surfaces for hexagonal crystals as a function of the anisotropy constants

(a) K_1 + ve $K_2 = 0$

(b) K_1 - ve $K_2 = 0$

(c) $K_2 = -K_1 = +ve$

TABLE 3

Expressions for magnetocrystalline energy for various crystal systems

Crystal Class	Typical Examples	Magnetocrystalline Energy	Definition of axes and direction of magnetization
Cubic	Fe, Ni	$K_1\left(\alpha_1^2\alpha_2^2 + \alpha_2^2\alpha_3^3 + \alpha_3^2\alpha_1^2\right) + K_2\alpha_1^2\alpha_2^2\alpha_3^2 +$	α_1, α_2, and α_3 are
	$MnFe_2O_4, Fe_3O_4$ (Spinel ferrites)	$K_3\left(\alpha_1^2\alpha_2^2 + \alpha_2^2\alpha_3^2 + \alpha_3^2\alpha_1^2\right)^2$	direction cosiness of M with
			respect to the cubic axes
	$Y_3Fe_5O_{12}$ (garnets)		of <100>
Hexagonal	Co	$K_1 \sin^2\theta + K_2 \sin^4\theta +$	
	$BaFe_{12}O_{19}$ (magneto pumbite structure)	$K_3 \sin^6\theta + K_4 \sin^6\theta \cos 6\phi$	
Trigonal	α-Fe_2O_3	$K_1 \sin^2\theta + K_2 \sin^4\theta + K_2^1 \sin^3\theta \cos\theta \cos 3\phi +$	The orientation of M is
	$FeBO_3$	$K_2 \sin^6\theta \cos 6\phi + K_3^1 \sin^6\theta +$	given by the usual
	ZnY ($BaZnFe_{16}O_{27}$)	$K_3 \sin^3\theta \cos^3\theta \cos 3\phi$	cylindrical coordinates θ, ϕ
Tetragonal	$CuCr_2O_4$	$K_1 \sin^2\theta + K_2 \sin^4\theta +$	
	Mn_3O_4	$K_3 \sin^4\theta \sin^2\phi \cos^2\phi$	
Orthorhombic	Fe_3O_4 -	$\sin^2\theta\left(K_1 \cos^2\phi + K_2 \sin^2\phi\right) +$	
	below 110 ^{o}K	$\sin^4\theta\left(K_3 \cos 2\theta + K_4 \sin^2\phi \cos^2\phi + K_5 \sin^4\phi\right)$	
	$GaFeO_3$	$\sin^2\theta \cos^2\theta\left(K_6 \cos^2\phi + K_7 \sin^2\phi\right)$	

However one must add a word of warning. The above expressions in Table 3 cover most of the cases of anisotropy, but for some materials, more complicated expressions will be required to fit the anisotropy; the effect of stress on the anisotropy should also be taken into account, and in some cases the anisotropy may be time dependent, and thus appear to vary with the method of measurement. Details of how to proceed when these complications are encountered will be given in the final sections of this chapter; the effect of stress is treated in Section 10 and the effect of relaxation is treated in Section 9. Finally it should always be remembered that the anisotropy constants are merely coefficients of a mathematical series expansion and have no direct physical meaning except in so far as they enable one to deduce the shape of the anisotropy energy surface.

4. Torque measurements

The method used to measure torque curves is basically simple. It is illustrated in Figure 9.

The sample is to be measured is usually a single crystal shaped in the form of a sphere or disc. It is suspended by means of a torsion wire in a magnetic field that is large enough to cause magnetic saturation. Note that since the sample is fully magnetized the torque should be independent of applied field. The torque on the sample is then given by the angular twist of the torsion wire. This can be read off directly or counterbalanced by rotating the torsion head until the crystal returns to its original position. Values of the torque for different directions of magnetization may be obtained by rotating the crystal and torsion head. Alternatively the electromagnet which provides the external magnetic field can be arranged to rotate around the crystal. Values of the torsion constant of suspension wire is known. Unless a sufficiently stiff torsion wire is used the system will possess certain unstable positions. Limiting torsion constants necessary for stability have been calculated by Harrison (1956)[3]. For cubic crystals, the maximum angular twist of the torsion wire θ_{max} is as follows: for the (100) plane

$$\theta_{max} < 0.25 \text{ radian}$$

(110) plane

$$\theta_{max} < 0.37 \text{ radian} \quad (K_1 + ve)$$
$$< 0.28 \text{ radian} \quad (K_1 - ve)$$

(111) plane

$$\theta_{max} < 0.16 \text{ radian}$$

For hexagonal crystals

Planes containing c-axis

$$\theta_{max} < 0.5 \text{ radian}$$

The instruments used for these measurements are known as torque magnetometers and the choice of instrument for any particular measurement depends on the following factors.

4.1 Physical dimensions of magnetic sample

Single crystals are essential if measurements of the intrinsic magnetocrystalline anisotropy are required. The preparation of suitable samples is sometimes not straightforward and in some cases only very small samples can be prepared. If measurements are required in several crystallographic planes it is usually preferable that all such orientations are possible with the same sample. For many purposes, spherical samples are found to be most convenient, and in the case of magnetic oxide crystals can be prepared accurately with diameters ranging from 5 mm to 0.25 mm. Details of the methods are given in Section 11. However for certain metals in which single crystals are most conveniently prepared in sheet form, a disc is a more suitable sample shape. The size and shape of the sample dictates the sensitivity and also the dimensions of the torque magnetometer. The following nomogram in Figure 10 gives an approximate idea of the torque range required as a function of sample volume and probable anisotropy. The regions enclosed by the series of straight lines correspond to the torque ranges

1 - 10^{-5} - 10^{-3} Nm (10^{2} - 10^{4} dyne cm)
2 - 10^{-7} - 10^{-5} Nm (1 - 10^{2} dyne cm)
3 - 10^{-9} - 10^{-7} Nm (10^{-2} - 1 dyne cm)
4 - 10^{-11} - 10^{-9} Nm (10^{-4} - 10^{-2} dyne cm)

In practice one can vary the volume of sample slightly to suit the

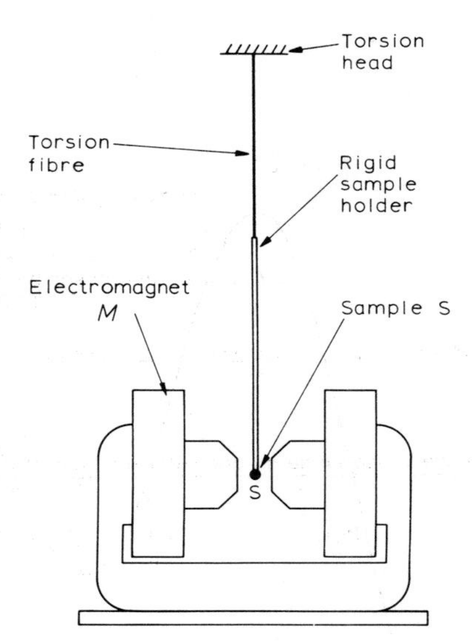

Figure 9 Schematic diagram showing the use of a torque magnetometer to measure anisotropy

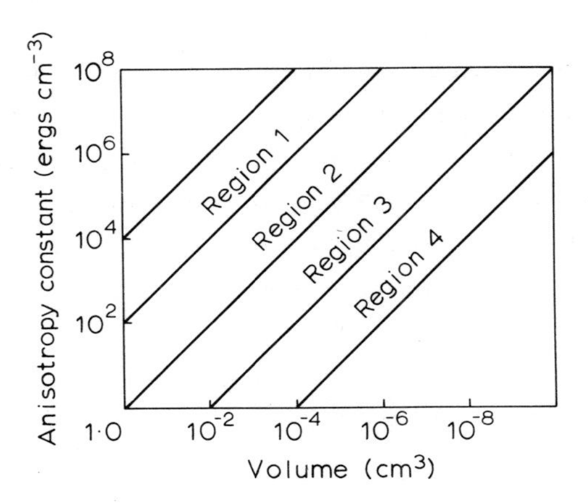

Figure 10 Torque range as a function of sample volume and approximate anisotropy constant. The regions enclosed by the straight lines correspond to the ranges

1 : 10^{-3} - 10^{-5} Nm 10^4 - 10^2 dyne cm
2 : 10^{-5} - 10^{-7} Nm 10^2 - 1 dyne cm
3 : 10^{-7} - 10^{-9} Nm 1 - 10^{-2} dyne cm
4 : 10^{-9} - 10^{-1} Nm 10^{-2} - 10^{-4}dyne cm

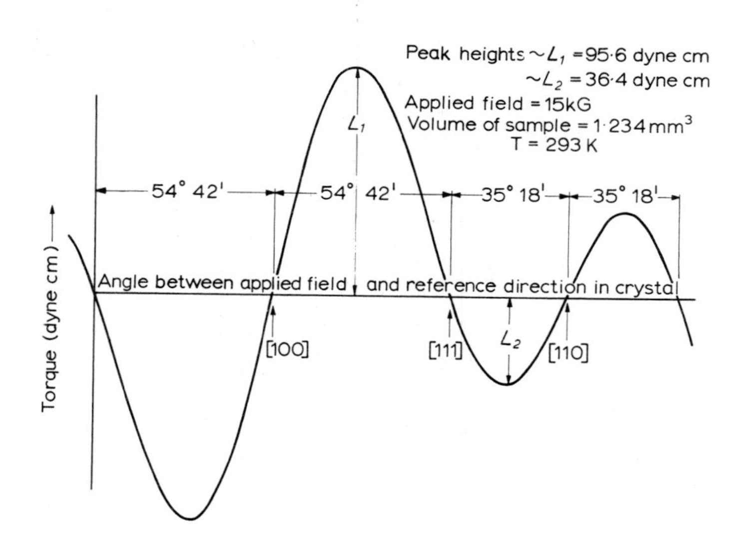

Figure 11 Experimentally recorded torque curve for a single crystal sphere of Fe_3O_4 mounted such that the field is applied in the (110) plane. Applied field = 1.5T=15kG. Temperature = 293°K. Peak heights L_1 = 956μ Nm (95.6 dyne cm), L_1 = 3.64μ Nm (36.4 dyne cm) volume of sample 1.234 mm^3

sensitivity of the available apparatus.

For example in studies of the anisotropy of the iron garnets at temperatures from 300 ^{O}K to 4.2 ^{O}K where the anisotropy is in the range 10^3 to 10^5 Jm^{-3} (10^4 to 10^6 erg cm^{-3}) using samples of volume 10^{-3} to 10^{-4} cm^3, an instrument with sensitivity in the range 10^{-7} to 10^{-9}Nm (1 - 100 dyne cm) has been found to be suitable. Note that if higher temperature measurements had been necessary (for temperature 300 - 500 ^{O}K where the anisotropy varies from 10^4 to 10) an instrument of higher sensitivity , i.e. 10^{-9} to 10^{-7} Nm (10^{-2} to 1 dyne cm) would have been necessary.

4.2 Magnetic field requirements

If the samples have a high magnetic anisotropy, and consequently require a large magnetic field for saturation, then either a large electromagnet or a very small magnetic gap is required to produce this field uniformly at the sample. Thus the torque magnetometer may have to be compact enough for the suspension to fit into a small magnet gap, and if measurements are required over a wide temperature range, the efffective gap available may be further reduced due to the insertion of dewar flasks and temperature control equipment. In practice although it is possible to design magnetometers and associated cryostats which will operate in gaps of 2.5 cm, experimental difficulties are considerably reduced if a gap of 5 cm is available. A typical electromagnet with a pole face diameter of 25 cm will produce fields of 1.5 T (15 kG) in such gaps. These fields are adequate for torque measurements on most materials. Higher fields can be obtained using superconducting or high current solenoids. If these fields are still lower than the anisotropy field for the material (which can be derived for cubic crystals as shown in Table 4) it is probable that torque measurements are not the easiest method of determining the anisotropy. A description of a more suitable method is given in Section 8. If measurements are required only in the region around room temperature, and no elaborate cryostats are required then smaller electromagnets can be used, if sufficient field uniformity around the sample position is obtained.

If measurements are required over a range of temperatures, and it is not possible to provide a method of stable temperature control, it may not be convenient to record the torque curves on a point by point basis. Many workers have found it convenient to develop automatic torque magnetometers in which some electrical means is used to counterbalance the torque on the sample. The torque magnetometer is then automatically balanced by some suitable electrical feedback network and the torque may be recorded directly as a function of angle of rotation of the magnet supplying the external magnetic field on the sample. (see Section 4.3).

TABLE 4

Minimum field required for saturation in cubic crystals

K_1 + ve		
(111)	B >	$4(3K_1 + K_2)/9M_s$
(110)	B >	$2K_1/M_s$
K_1 - ve		
(100)	B >	$-2K_1/M_s$
(110)	B >	$-(2K_1 + K_2)/2M_s$

A typical experimental curve for a single crystal of magnetite (Fe_3O_4), approximately 1.33 mm diameter and mounted in the (110) plane is shown in Figure 11. For this sample a torque magnetometer of sensitivity in the range 10^{-7} - 10^{-5} Nm(1-100 dyne cm) was used. In the following section, experimental details of magnetometers to cover a range of measurements for 10^{-13} to 10^{-2} Nm (10^{-6} to 10^{5} dyne cm) are described. Obviously the choice of instrument is quite dependent on the torque range to be measured.

4.3 Details of the experimental magnetometers

The range of instruments developed for torque measurements is summarised in Table 5.

In the torque range 10^{-5} to 10^{-3} Nm (10^{2} - 10^{4} dyne cm), the type developed by Penoyer[4] has been the model for many later developments. In this instrument, schematically shown in Figure 12, lateral displacement of the sample is avoided by the use of bearings which allow only rotational motion.

These bearings determine the lower limit on the torque measured by

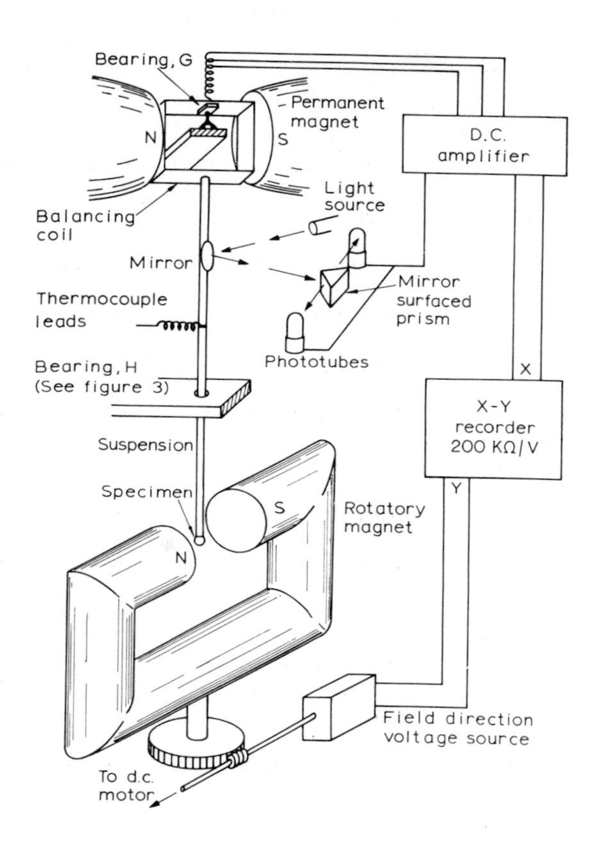

Figure 12 Schematic diagram of Penoyer magnetomer.[4]

(Reproduced by permission of the American Institute of Physics)

TABLE 5

List of torque ranges covered by experimental magnetometers

Torque Range 10^{-7} Nm dyne cm	Sensitivity 10^{-7} Nm dyne cm	Application	Reference
$1-10^{-6}$	10^{-6}	Thin magnetic films	Humphrey et al (1963)[19]
$10^{-2}-10^{-4}$	10^{-4}	Thin films	King et al (1964)[18]
$1-10^{-3}$	10^{-3}	De Haas v Alphen	Condon et al (1964)[16]
$1-10^{-3}$		Ferrites	Maxim (1969)[17]
$1-10^{-2}$	10^{-3}	Thin films	Bransky et al (1968)[13]
$10-10^{-2}$		de Haas v Alphen	Croft et al (1955)[14]
$3-5\times10^{-2}$	.05	Nickel films	Westwood et al (1970)[15]
$10^{-1}-10^{3}$		Ferrites	Fletcher et al (1969)[11]
$10-10^{3}$	1	Nickel disc	Birss et al (1963)[8,9]
$10^{2}-10^{4}$		-	Aldenkamp et al (1960)[5]
$10^{2}-10^{5}$	1	Ferrite crystal measurements K_1K_2	Penoyer (1959)[4]
$10^{2}-10^{5}$	10	Nichel sphere	Aubert (1968)[6]

this instrument. The upper bearing which supports the entire weight of the suspension consists of a steel needle on sapphire. The lower bearing may be made from a non-magnetic rod in a sapphire ring bearing, but Penoyer also used an air bearing as shown in Figure 13. However this causes experimental difficulties due to the flow of air through the bearing if torque measurements are required in controlled atmospheres. For measurements at low temperatures the author has successfully used a sapphire ring lower bearing, which could operate in liquid helium if necessary.

In order to achieve the high balancing torques of up to 10^{-2} Nm (10^5 dyne cm), and without requiring a large number of ampere turns on the solenoid a magnetron magnet is used to supply the static field, i.e. 0.2T (2000 Oe) in a 5 cm gap. Variations in the sensitivity of the balance are obtained by using different tappings (i.e. 5 turns, 50 turns) on the balancing coil.

An instrument capable of torque measurements up to 10^{-2}Nm (10^5 dyne cm), but which avoids the use of bearings was developed by Aldenkamp, Marks and Zijlstra[5] (1960). The sample is rotated in a magnetic field by a vertical motor driven shaft. The shaft incorporates a transducer, shown in Figure 14 which is highly sensitive to twisting but shows great resistance to bending. Thus in this arrangement, the shaft does not need a bearing to prevent transverse displacements. The transducer is made from two circular disks which are attached to each other by flat springs. The displacement of the 2 circular disks relative to each other is measured by an inductive displacement meter connected to a direct reading measuring bridge. The sensitivity of this arrangement is such that a displacement of a few micrometers corresponds to full scale deflection. Torque curves are recorded by automatically rotating the samples in a magnetic field by the motor driven shaft and recording the output of the displacement meter as a function of rotation angle. The instrument sensitivity can be varied from 150 to 1000 dyne cm per cm deflection of output recorder. The mechanical construction ensures that the instrument is very robust in operation, but attempts to increase the sensitivity by electronic amplification are limited by elastic after effect in the transducer.

The next magnetometer to be discussed in this range, is worthy of special attention. In the instrument developed by Aubert[6] (1968) for

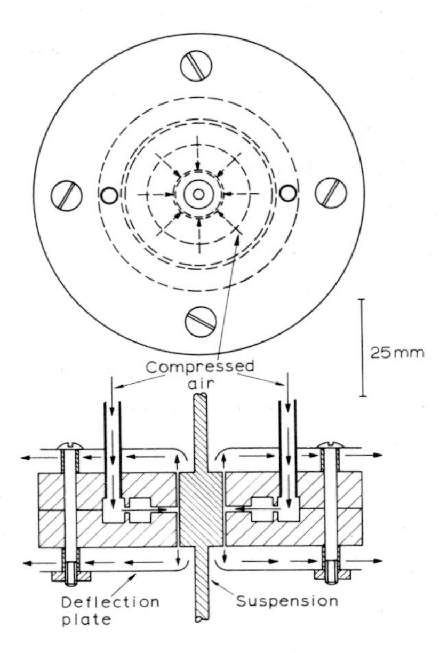

Figure 13 Air bearing.

(Reproduced by permission of the American Institute of Physics)

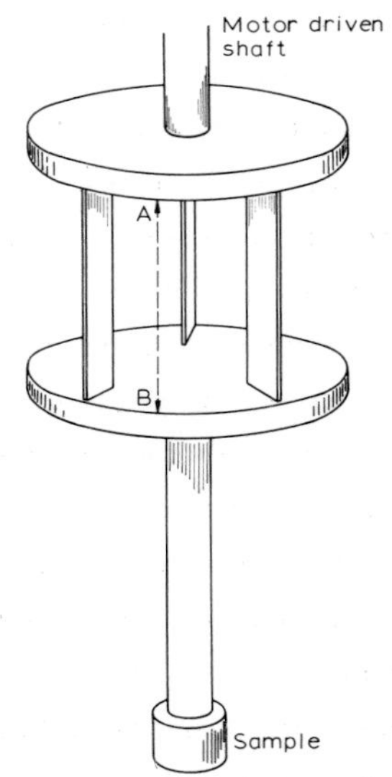

Figure 14 Principle of torque transducer used in the Aldenkamp, Marks and Zijlstra magnetometer.[5]

(Reproduced by permission of the American Institute of Physics)

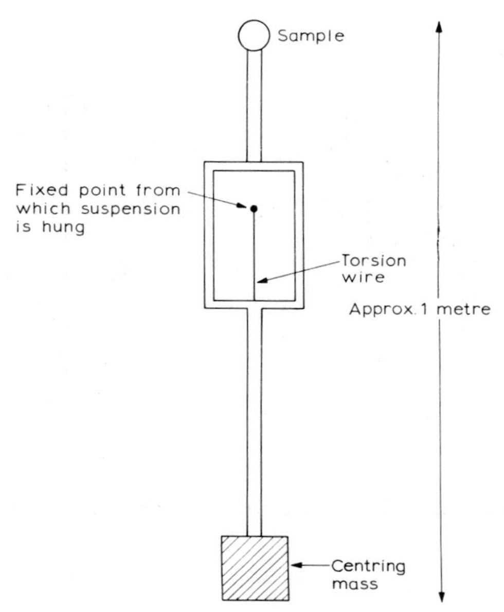

Figure 15 Schematic diagram of suspension of Aubert magnetometer[6]

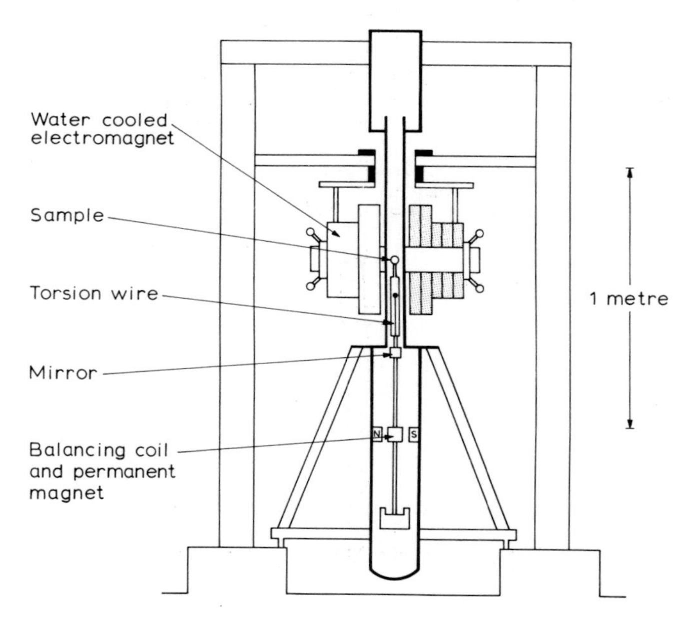

Figure 16 Detailed diagram of Aubert magnetometer in operational position.[6]

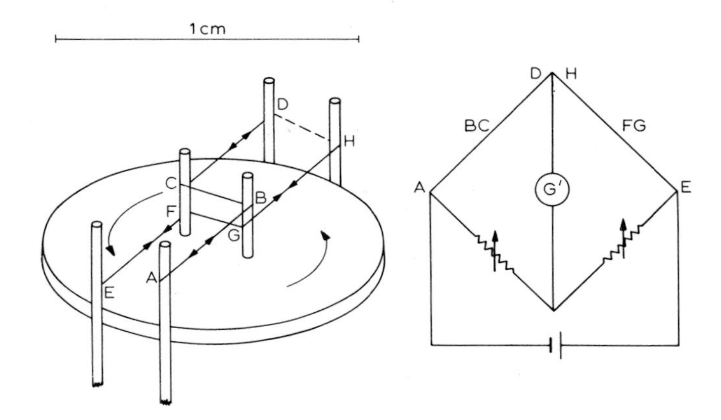

Figure 17 Diagram of principle of strain gauge torque meter.

(Reproduced by permission of the Institute of Physics)

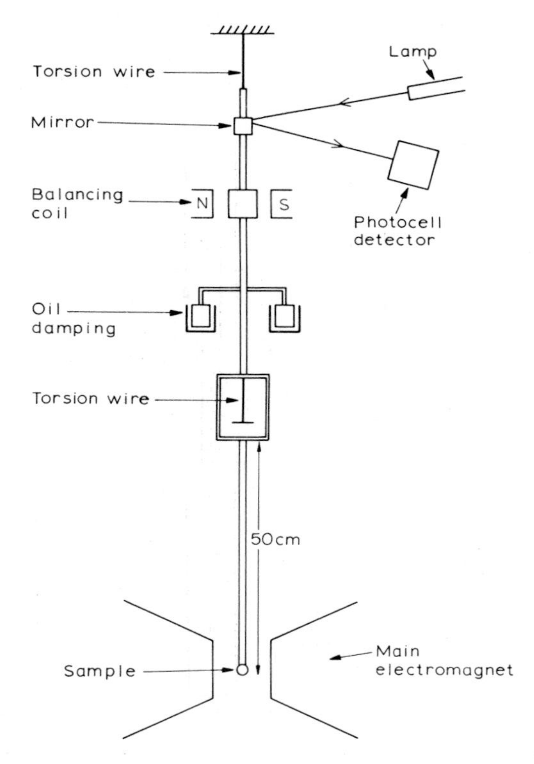

Figure 18 Schematic diagram of Gerber and Vilim magnetometer.[10]

(Reproduced by permission of the Institute of Physics)

extremely accurate measurements on single crystals of nickel, effects due to anisotropy of magnetization were also detected, and accuracies of some 10^{-7}Nm (dyne cm) in torque of 10^{-2}Nm (10^5 dyne cms) are obtained.

In the Aubert instrument, the sample is suspended from a torsion wire, and the sample is held centrally between the pole pieces of a large electromagnet under the action of a force provided by a heavy mass attached below the sample holder. The apparatus is shown schematically in Figure 15. The mass m is adjusted such that the restoring force is greater than the force due to the magnetic image of the sample in the pole pieces of the electromagnet.

The details of the apparatus are shown in more detail in Figure 16.

The movable electromagnet can be rotated to provide an applied field perpendicular to the vertical suspension axis. By means of optical detection and electronic feedback control, the torque exerted by the samples is balanced by sending the feedback current through a coil placed in a permanent magnet. Aubert states that the error deflection for the maximum measurable torque is less than 6 seconds of arc and the position of the electromagnet with respect to the sample is known to this accuracy for 36 equidistant positions, also by optical means. The maximum measurable torques are 10^{-2} Nm (10^5 dyne cm) for the high temperature apparatus (up to 1000 $^{\circ}$C) and 0.3 Nm (3.10^6 dyne cm) for the low temperature one (down to 2 $^{\circ}$K). The highest fields available are, respectively 1.2T (12 k Oe and 2T (20 kOe) and the temperature of the sample is regulated within 0.05 $^{\circ}$C. The feedback current is measured by potentiometric methods with a relative accuracy of 10^{-5}. A detailed description of this torque magnetometer and its application is given by Escudier (1975)[6].

An alternative way of avoiding bearing friction, which gives a positive location of the specimen in the magnetic field and insensitivity to unwanted translational forces and torques has been described by Birss and Wallis (1963)[8]. The torque on the sample in the range 0.1 to 10μ Nm (1 to 10^2 dyne cm), is arranged to produce a strain in a resistance wires ABCD and EFGH as shown in Figure 17. The wires form the two arms of a Wheatstone Bridge and the out of balance current in the detector G is proportional to the change in resistance and thus the torque on the crystal if the elastic limit on the wires is not exceeded.

The actual curves of torque against direction are produced by rotation of the external magnet around the sample. The method is deemed to be suitable for measurements over a wide range of temperatures and for measurements of anisotropy as a function of pressure. The pressure dependence of K_1 for Ni was determined using this method by Birss and Hegarty (1969)[9].

For a more sensitive magnetometer of the Penoyer type we have previously discussed, it is possible to dispense with the top bearing and suspend the balancing coil from a torsion fibre. However this leads to a loss in physical stability, and this may be counteracted by introducing a lower suspension. Magnetometers of this type have been described by Gerber and Vilim (1968)[10], by Fletcher et al (1969)[11] and by Pearson (1959)[12]. A schematic diagram of the Gerber and Vilim magnetometer is given in Figure 18.

As we go to higher sensitivities, i.e. the torque range 10^{-9} to 10^{-7} Nm (10^{-2} to 1 dyne cm) the magnetometer becomes more susceptible to extraneous influences, such as mechanical vibrations and magnetic impurities in the sample holder and magnetometer. Measurements at high fields and low temperatures become more difficult as a consequence and in general more care is required for torque measurements in this range.

An ingenious method of adapting a commercial electric recording microbalance into a sensitive torque magnetometer has been described by Bransky et al (1968)[13]. To measure a vertical torque, a Cahn RC electro-balance is rotated by 90° and coupled to the suspension on which the sample is mounted as shown in Figure 19. As the weighing loops are approximately 12 cm apart, and the rated sensitivity of the balance is 0.2 μg, the balance is capable of measuring torques of the order of 10^{-10} Nm (10^{-3} dyne cm).

Other freely suspended magnetometers which are suited for operation in this range are described by Croft, Donahoe and Love (1955)[14] and shown in detail in Figure 20. A somewhat similar instrument in which the deflection of the suspension is detected in inductive rather than photoelectric methods is described by Condon and Marcus (1961)[16]. Incidentially both instruments were originally developed for measurements of the de Haas-van Alphen effect. The latter instrument was adapted from a commercially available instrument, the Weston inductronic amplifier.

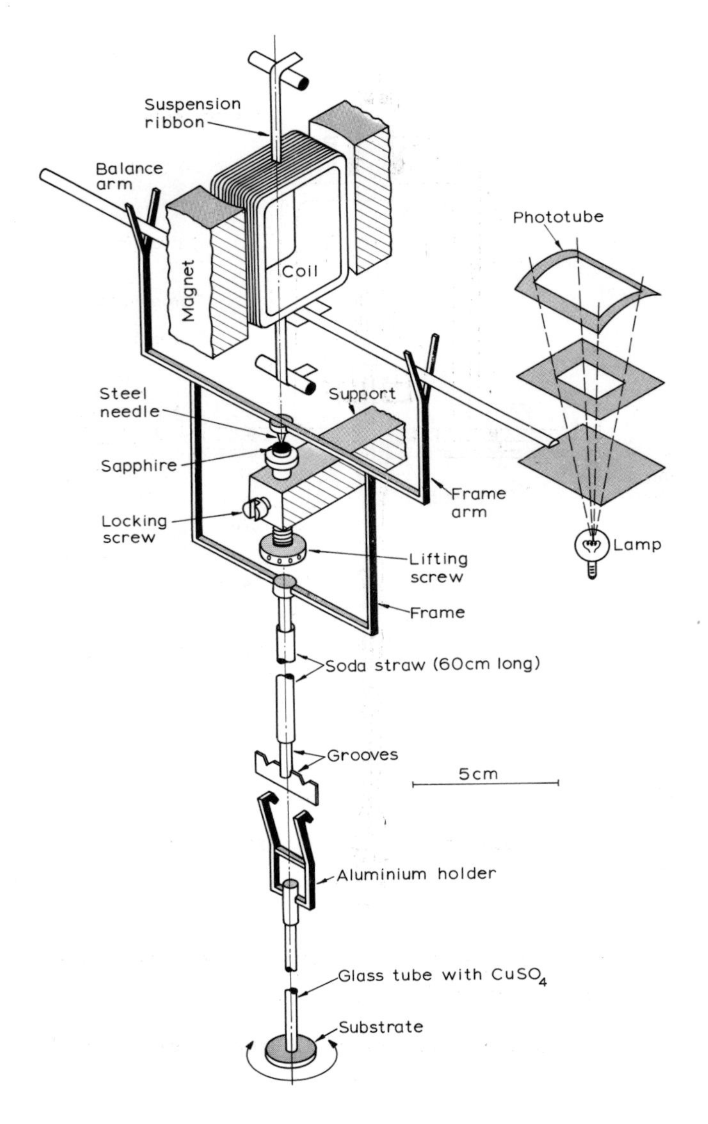

Figure 19 Conversion of microbalance into torque magnetometer as shown by Bransky, Hirsch and Bransky (1968)[13]

(Reproduced by permission of the Institute of Physics)

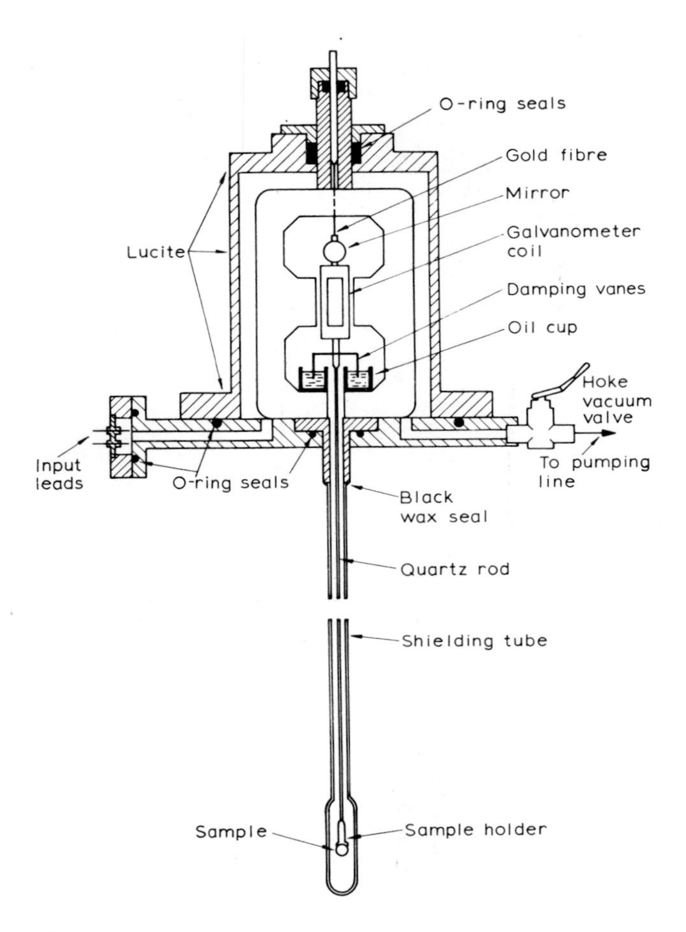

Figure 20 Diagram of suspended magnetometer due to Croft, Donahoe and Love (1955)[14]

(Reproduced by permission of the American Institute of Physics)

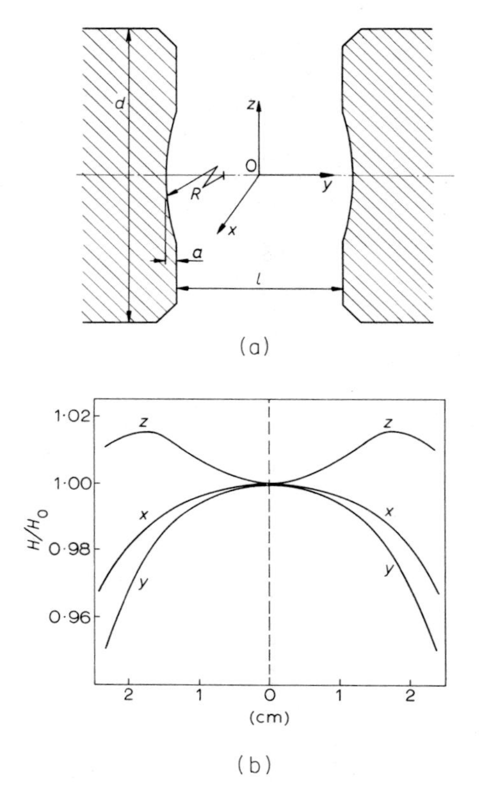

Figure 21 The use of shaped pole-pieces to produce a field configuration that gives a stable position for a magnetic sample placed at the centre.

(Reproduced by permission of the Institute of Physics)

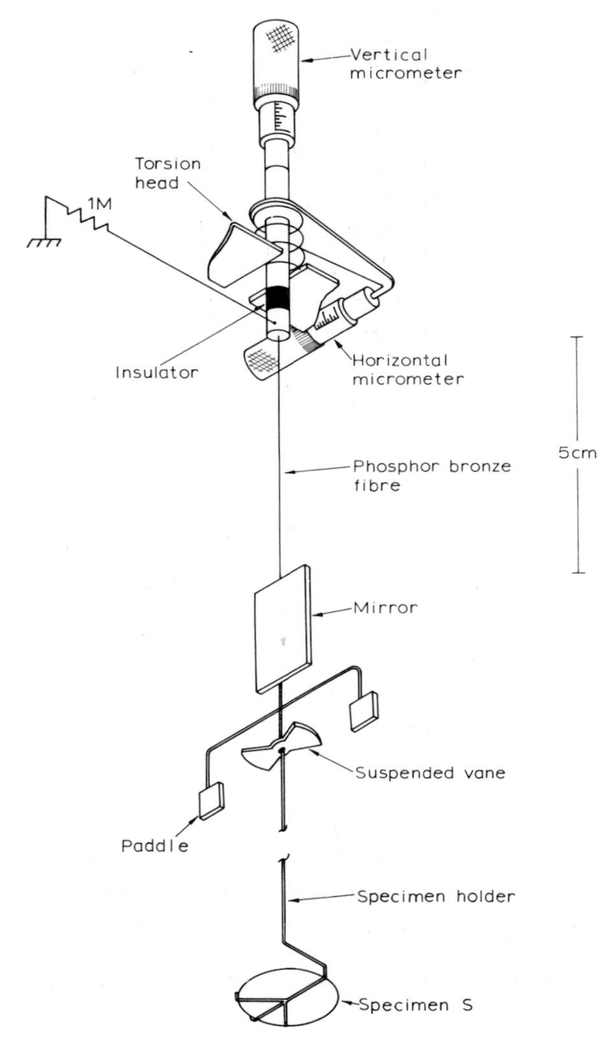

Figure 22 Schematic diagram of suspension of electrostatic torque magnetometer.

(Reproduced by permission of the Institute of Physics)

In all the instruments described in this section, care must be taken to place the sample as near as possible to the centre of the magnet gap, and the field uniformity in the gap should be as high as possible to prevent the appearance of parasitic torques.

An ingenious way of centring the sample in the gap is given by Maxim (1965)[17]. By shaping the pole pieces of the electromagnet around the sample, as shown in Figure 21 then a maximum in the field can be produced at the gap centre. A ferromagnetic sample placed at this point is in stable equilibrium with respect to displacements from the rotation plane. The curves shown are for the case where d = 10 cm, ℓ = 5.5 cm, a = 3.5 mm, R = 6 cm.

For magnetometers in the torque range 10^{-9} - 10^{-11} Nm (10^{-2} to 10^{-4} dyne cm) in the highest sensitivity range, special precautions must be observed.

King, Robinson, Cundall and Hight (1964)[18] used a magnetometer in which the balancing was done electrostatically. Thus they were able to avoid the use of current leads attached to the instrument suspension, which possessed no external ligaments except for the suspension fibre, as shown in Figure 22. Torques up to 10^{-9} Nm (10^{-2} dyne cm) could be produced by electrostatic means and sensitivities of 10^{-11} Nm (10^{-4} dyne cm) were achieved. Rotational damping of the system was obtained by means of small aluminium paddles which were immersed in a paraffin mixture.

Finally we come to one of the most sensitive instruments developed by Humphrey and Johnston (1963)[18] which will measure torques down to 10^{-13}Nm (10^{-6} dyne cm). The torque movement is shown in Figure 23. The torque meter movement, torsion fibres and fixed supporting and tensioning members have all fabricated from fused silica to provide an integral mechanical stable assembly. The fused silica suspension is gold plated to make it conducting. The torque balance is provided by the magnetic field from external permanent magnets acting on the current flowing through the straight element parallel to the torsion axis. Twenty milliamperes balancing current corresponded to a maximum torque of 7.0 x 10^{-7}Nm (7.0 dyne cm). For fuller experimental details in the construction, readers are referred to the original paper[18]. The instrument was designed to measure anisotropy in thin films, and the authors point out that in measuring very small magnetic torques, many difficulties may arise. For example considerable stray torques can be introduced by the sample substrate.

The performance of this instrument is extremely impressive with drift rates as low as 0.4 Nm per hour (4 x 10^{-6} dyne cm per hour). It has obviously benefited considerably from development carried out on the instrument in its original form as an accelerometer and it may not be so easy to reproduce its performance in other laboratories. The torque meter was calibrated as a mass balance using a known weight.

The torque magnetometers described so far have been built for use in conventional electromagnets with the sample suspended perpendicular to the field direction. Several static magnetometers have been described which are designed for operation in high field solenoids (Vanderkooy (1969)[20], Voigt and Foner (1971)[21]. In the instrument designed by Voigt and Foner, shown schematically in Figure 24, the torque on the sample is compensated by a current applied to a solenoid on which the sample is attached. Rotation of the sample is detected by a capacitator method, Be-Cu wires of 3 mm length and thickness 0.1 to 0.25 mm were used for the torsion wire and the whole arrangement is rotated with the solenoid by a worm gear drive. A torque range up to 10^{-3}Nm (10^{4} dyne cm) can be measured, and an accuracy measurement of 2% in torques of 10^{-5}Nm (100 dyne cm) is possible.

4.4 Errors in experimental measurements

In measurements on low anisotropy materials, it is essential to remove extraneous torques present. Sources of error in torque measurements may be related to the following factors.

(1) The position of the sample in the magnet.

(2) Shape effects in sample

(3) Impurities in suspension or sample holder.

(4) Instrumental errors.

In practice they may be eliminated by a systematic procedure as follows. Make a complete torque measurement, at the appropriate applied magnetic field, but without a sample present. The torque curve should be a horizontal straight line, which is also reversible.

Any torques produced are due to

(a) magnetic impurities in the suspension or sample holder

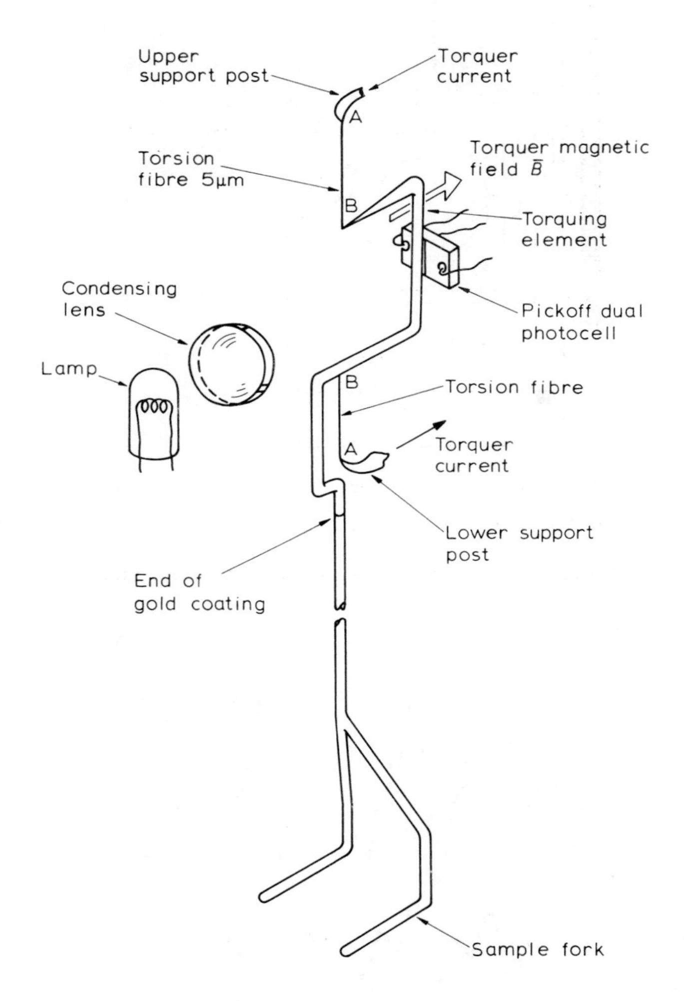

Figure 23 Torque meter movement of the Humphrey and Johnston magnetometer.[19]

(Reproduced by permission of the American Institute of Physics)

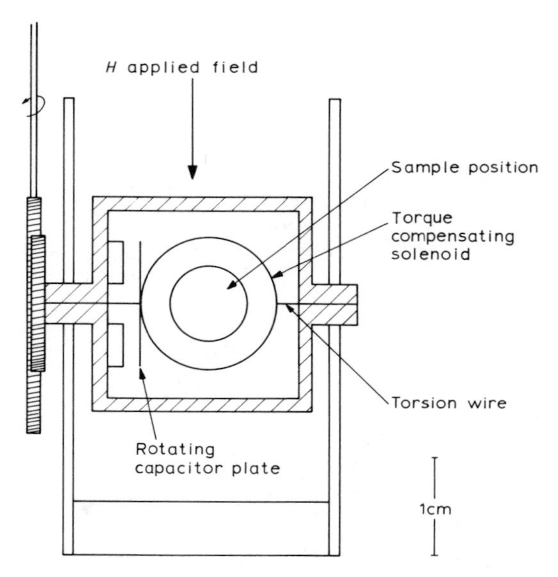

Figure 24 Schematic view of high field solenoid torque meter due to Voigt and Foner (1971)[21]

(Reproduced by permission of the Institute of Physics)

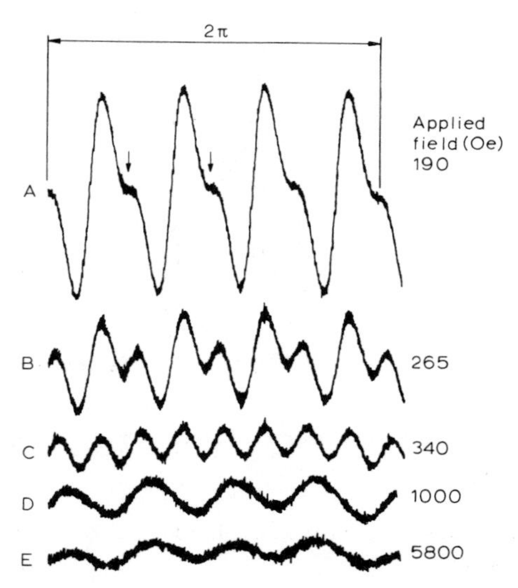

Figure 25 Torque curves exhibited by a square block as a function of magnetic field.

(Reproduced by permission of the American Institute of Physics)

(b) interaction between the applied magnetic field and any part of the torque balancing circuit that is sensitive to external magnetic fields.

Next the sample should be replaced and the torque curve remeasured. Parasitic effects of period π and 2π may be observed now. These are due to the following factors:

(1) Position of the sample in the magnet

(a) In the case of a uniform magnetic field, the sample is not central. This will produce a torque due to the image effect in the pole pieces.

(b) The sample is placed in a non uniform magnetic field.

(c) The centre of the sample is not coincident with the axis of the magnetometer.

When both (b) and (c) are present, then considerable additional torques may develop.

For the case of a disc in field gradient, $\frac{dB}{dx} = 10^{-2}Tm^{-1}$ (1 oe cm^{-1}). $M_s = 1000$ k Am^{-1} (emu) with disc radius 0.5 cm, when the suspension axis and centre are one radius length apart.

The error torque is of the order of 50 J m^{-3} (500 erg/cm^3) per unit volume. If the torque does not repeat in 180°, then it usually means that the magnetometer is not adequately centred in the magnet.

In the case of disc shaped sample, an additional error is introduced if the disc is not mounted horizontally in the field. Here the demagnetization energy of the disc tends to keep the magnetization lying in the plane and thus introduces an additional torque of the form

$$L = -M_s B \quad (1 - \cos \alpha_o)\sin 2\theta \qquad (26)$$

where α_o is the angle between the plane of the disc and the magnetic field.

Note if $M_s = 1000$ k Am^{-1} (emu), B = 0.5T (5000 oe), $\alpha_o = 1^o$, a maximum torque of 100 J/m^3 (1000 erg/cm^{-3}) would be introduced, which would vary linearly with applied field, and have a sin 2θ type variation.

(2) Shape of the sample

For measurements of high accuracy it is necessary to know the additional torques introduced by shape anisotropy of the sample. As an example let us consider a sample nominally shaped in the form of a sphere, which has an elongation of a diameter by amount ε to produce an ellipsoidal sample. Knowing the demagnetization factors of ellipsoids we find for the torque/unit volume.

$$L = \frac{\mu_o M_s^2 \varepsilon \sin 2\phi}{5} \tag{27}$$

For a nickel sample, this is of the order of $6\varepsilon \times 10^4$ Nm/m^3 ($6\varepsilon \cdot 10^5$ dyne cm/cm^3) at room temperature which is to be compared with the value due to the magnetocrystalline anisotropy of 3×10^3 Jm^{-3} (3×10^4 erg cm^{-3}). Thus for measurements of 0.1% accuracy, it is necessary to form spheres in which ε is less than 5×10^{-5}. Fortunately as will be seen in a later section it is possible to produce spheres of accuracy about a micron, such that the above requirements can be fulfilled with samples of 8 or 9 mm diameter.

An interesting feature of the shape anisotropy contribution is that if the sample is placed in a saturating field, such that $B \gg \mu_o M_s$ and we can assume that the magnetization is uniform, then for a sample of arbitrary shape, the magnetostatic energy contains harmonic terms depending only on the second power of the magnetization direction cosines. Thus the shape anisotropy of the sample will contribute only terms in $\sin 2\phi$ to the torque where ϕ is the direction of the magnetization relative to a reference direction in the sample.

Note that small internal defects in samples, i.e. holes, inclusions, etc., which produce local demagnetizing fields will accordingly introduce a $\sin 2\phi$ component in the torque.

For fields lower than $\mu_o M_s$. and where the magnetization in the sample is non uniform quite different contributions to the torque can arise. An interesting demonstration of this has been given by Hagedorn and Gyorgy (1968)[22]. Measurements on thin plates cut in the form of equilateral triangles or squares give torque curves of $\sin 6\theta$ and $\sin 4\theta$ respectively. The torque curves decrease in amplitude as the applied field is increased. Although we would expect these shapes to be isotropic when $B > \frac{\mu_o M_s}{3}$ for uniform magnetization causing the torques to go to zero, appreciable anisotropy remains for applied fields up to 15 B_s in some cases for square plates. This implies that the magnetization is still non uniform in the sample up to these fields, and emphasizes the importance of using samples such as ellipsoids which are uniformly magnetized in low fields. An example of the torque curves exhibited by a square cross section block, $M_s = 30\ kAm^{-1}$ (emu) as a function of the applied field is given in Figure 25.

(3) Impurities in the suspension or sample holder

Although these effects can be allowed for by measurements with and without the sample, a careful choice of sample holder will reduce the correction required. For low temperature measurements a suitable plastic is methyl methacrylate but in cases where mechanical strength is also required a fibre bonded plastic has been used. Care must be taken in the machining of the sample holder that no pick up of ferromagnetic material takes place and thorough cleaning takes place after the machining process.

(4) Instrumental errors

Under this heading we classify errors due to the suspension of the instrument, such as sticking caused by excessive oil damping (where present), bearing friction, torques contributed by current ligaments connected to the balancing coil, mechanical hysteresis in the suspension . These are always present and usually can be detected by measurements made in the absence of a sample.

4.5 Calibration methods

A major difficulty in torque measurements is the calibration of the apparatus in absolute terms. In this section convenient methods of calibration for different torque ranges are given. Note that calibration is not only for giving the magnitude but the sign of the torque and thus the sign of the anisotropy constants of the sample under investigation. The most convenient method of calibration is the use of a standard sample. In the torque range $10^{-8} - 10^{-6}$ Nm (0.1 to 10 dyne cm) a single crystal sphere of yttrium iron garnet (YIG) is a useful standard, which may be used from room temperature down to 4.2K. A suitable orientation is the (100) plane which will produce a torque curve of form $L = \frac{K_1}{2} \sin 4\theta$ at room temperature in applied fields over 0.15T (1500 oersteds). The accuracy of the shape of sphere largely determines the accuracy of the calibration as discussed in a previous section. Values of K_1 may also be derived from accurate microwave resonance experiments as an independent check on the standard used.

For diameters , 0.5 mm and 2.5 mm diameter respectively the torque amplitudes at room temperature are 0.203 and 25.4×10^{-7} Nm (dyne cm) respectively. Values of K_1 from 500 to 1.5 K are given in Figure 26. A method relying on the calculated shape anisotropy of a standard sample has been used by Aldenkamp, Marks and Zijlstra[5] for calibration of torques up to 10^{-3} Nm (10,000 dyne cm) and this is considered suitable for torques from 10^{-4} - 10^{-3} Nm (10,000 dyne cm). They used a rod shaped sample of pure nickel. Thus in a strong magnetic field, the torque is given by

$$L = \tfrac{1}{2}\mu_0(D_\perp - D_\parallel)M^2 V \sin 2\theta \tag{28}$$

$$[L = \tfrac{1}{2}(N_\perp - N_\parallel)M^2 V \sin 2\theta]_{\text{emu}}$$

where $D_\perp$ = the demagnetizing factor perpendicular to the bar axis

$D_\parallel$ = the demagnetizing factor parallel to the bar axis

M = saturation magnetization

V = volume of bar

θ = angle between magnetization at axis of bar

Note that in automatic torque magnetometers, the torque is plotted as a function of the angle made by the applied magnetic field (not the magnetization) with a reference direction in the samples.
For a nickel cylinder, length 17 mm, diameter 1 mm, $(D_\perp - D_\parallel) = 0.5$ (see standard demagnetization factors of cylinders) $V = 1.33 \times 10^{-8}$ m^3
Taking $M_s = 485$ k Am^{-1} (emu).

This gives $L_{max} = 0.94 \times 10^{-3}$ Nm (9400 dyne cm).
If a check on the linearity of the magnetometer is required, it is convenient to depend on a more direct method of calibration. A convenient direct method of calibration is to mount a small solenoid on the sample holder , and vary the current passing through the solenoid. When a field is applied, the solenoid experiences a torque proportional to the effective area, the number of turns, the current passing through the solenoid. When a field is applied, the solenoid experiences a torque

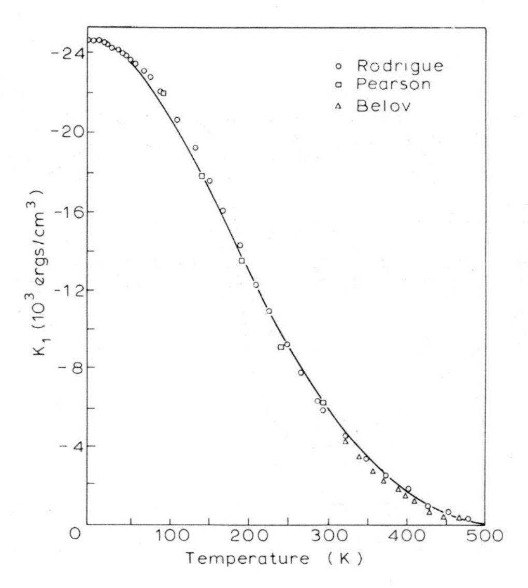

Figure 26 Values of K_1 for yttrium iron garnet (YIG) between $350^{\circ}K$ and $1.5^{\circ}K$.

(Reproduced by permission of Academic Press Inc.)

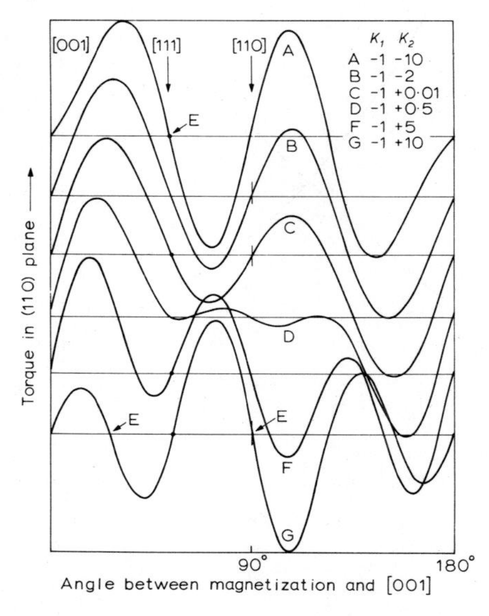

Figure 27 Torque curves in the (110) plane for different values of K_1/K_2.

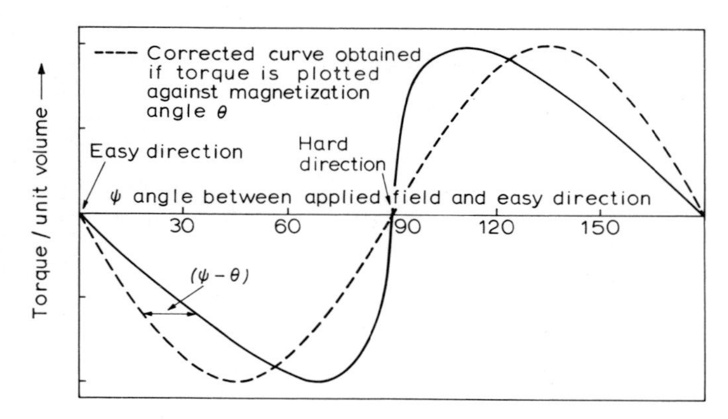

Figure 28 Torque plotted against field direction showing distortion introduced by the magnetization lagging behind the field.

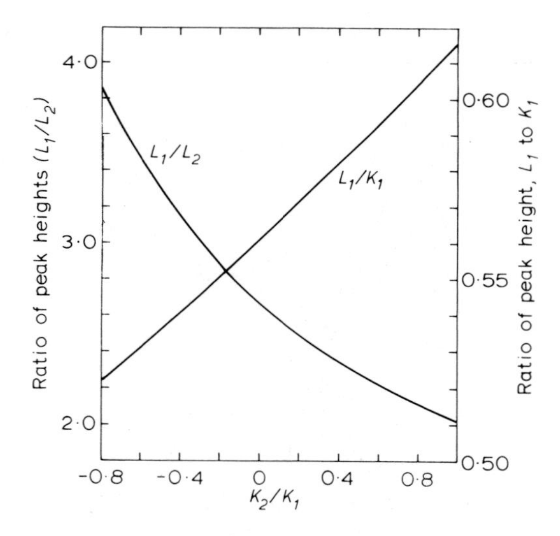

Figure 29 Ratio of peak heights of (110) torque curves L_1/L_2 versus K_2/K_1, and L_1/K_1 versus K_2/K_1.

proportional to the effective area, the number of turns, the current passing through it, and the applied magnetic field. The only disadvantage with this method is that care must be taken to avoid spurious torque associated with magnetic and mechanical forces on the wires leading to the solenoid. Finally in very sensitive instruments, the usual method is to calibrate the torsion wire of the suspension, in an independent experiment.

4.6 Analysis of experimental results

Let us consider the case which is most usually encountered that is the derivation of the anisotropy constants K_1 and K_2 from torque measurements on a sample with cubic symmetry. The expression for the torque curve due to the intrinsic magnetocrystalline anisotropy will depend on the crystallographic plane in which the torque is measured, examples calculated for curves in the principal planes are given in Table 6.

Torque expressions for cubic crystals in different planes

Note the simplest expression which gives both K_1 and K_2 is given by torque curves in the (110) plane, and the type of curve which is obtained for different ratios of K_1 and K_2 is given in Figure 27.

Note the easy directions are marked by E → for curves A and F and are found as specified in Section 2.

If K_1 only is required, measurements need by made only in the (100) plane where the anisotropy constant K_1 is given by the (torque curve peak height) x 2.

However for more complicated curves (i.e. (110)) the first stage in the treatment of the experimental results is to convert the experimental curves of torque versus angle of applied field to curves of torque against magnetization direction. It can often happen that the magnetization direction may lag considerably behind the magnetic field direction. The effect of this is to produce experimental torque curves which are distorted as shown in Figure 28. However the experimental curve can be corrected in the following way, if the sample magnetization and applied field are known.

$$\text{Torque} \quad \tau = MB \sin(\psi - \theta)$$

TABLE 6

Values of $dE_k/d\theta$ as $f(\theta)$
(θ is measured in $(hk\ell)$ plane from direction $[h_0k_0\ell_0]$)

$(hk\ell)$	$[h_0k_0\ell_0]$	$dE_k/d\theta$
100	001	$K_1(\sin 4\theta)/2$
100	011	$K_1(-\sin 4\theta)/2$
110	001	$K_1(2\sin 2\theta + 3\sin 4\theta)/8 + K_2(\sin 2\theta +$ $4\sin 2\theta + 4\sin 4\theta - 3\sin 6\theta)/64$
110	$1\bar{1}0$	$K_1(-2\sin 2\theta + 3\sin 4\theta)/8$ $+ K_2(-\sin 2\theta + 4\sin 4\theta + 3\sin 6\theta)/64$
110	$1\bar{1}1$	$K_1(-2\sin 2\theta - 7\sin 4\theta)/24 + K_1(\cos 2\theta$ $- \cos 4\theta)/3\sqrt{2} + K_2(-3\sin 2\theta - 28\sin 4\theta$ $- 23\sin 6\theta)/576 + K_2(3\cos 2\theta - 8\cos 4\theta$ $+ 5\cos 6\theta)/(144\sqrt{2})$
111	$1\bar{1}0$	$K_2(\sin 6\theta)/18$
111	112	$K_2(-\sin 6\theta)/18$
211	$01\bar{1}$	$K_1(2\sin 2\theta - 7\sin 4\theta)/24$ $- K_2(-13\sin 2\theta + 20\sin 4\theta - 25\sin 6\theta)/576$
211	$1\bar{1}\bar{1}$	$K_1(-2\sin 2\theta - 7\sin 4\theta)/24$ $- K_2(13\sin 2\theta + 20\sin 4\theta + 25\sin 6\theta)/576$

where $\psi - \theta$ is angle between field and magnetization

$$\therefore \quad \overline{\psi - \theta} = \sin^{-1} (\tau/MB) \tag{29}$$

$(\psi - \theta)$ can be computed for each point of the experimental torque curve which can then be replotted against magnetization direction. Obviously if B is very large the corrections are very small and can be neglected.

It was pointed out by Shenker (1955)[23] that the heights of the maxima in the torque curves are unchanged by this correction, and he showed how K_1, K_2 could be derived from the peak heights of (110) torque curve. This method is very useful for a rapid determination of the anisotropy coefficients. However the method assumes the theoretical expression for the torque curve is correct, i.e. that only terms corresponding to the intrinsic magnetocrystalline anisotropy are present.

Bearing these points in mind, the Shenker method for (110) plane measurements on cubic crystals procceds as follows,

(a) Measure the heights L_1, L_2 of the peaks of the experimental torque curve as shown for example in Figure 11.

(b) Read off the value of $\frac{K_2}{K_1}$ corresponding to the ratio L_1/L_2 in Figure 29. For values outside the range shown in this figure values may be interpolated from the Table 7.

(c) From the graph of L_1/K_2 against K_2/K_1 in Figure 29, L_1/K_1 may now be derived. As we know L , K may be deduced. From which we may derive K_2 as the ratio K_2/K_1 is already known.

In cases where it is uncertain whether the anisotropy can be fitted simply by K_1 and K_2, a more general analysis of the results is required, and the following method may be used.

(1) Measure the torque L_ψ at equal intervals of ψ the angle made by the applied field (usually 10°) intervals in range $0 - 180^\circ$).

(2) For each value of L_ψ calculate the angular difference between the magnetization and field direction ($\psi - \theta = \sin^{-1}(L/MB)$).

(3) Replot L_ψ versus θ the magnetization direction.

(4) Derive a set of values of L for equal intervals of θ

TABLE 7

Ratio of peak heights of torque curve in (110) plane versus K_2/K_1 and the ratio of L_1 to K_1 versus K_2/K_1

K_2/K_1	L_1/L_2	L_1/K_1	K_2/K_1	L_1/L_2	L_1/K_1
-2.0	14.704	0.4715	1.8	1.749	0.6654
-1.8	10.286	0.4791	2.0	1.698	0.6781
-1.6	7.748	0.4871	2.2	1.652	0.6910
-1.4	6.174	0.4954	2.4	1.610	0.7041
-1.2	5.131	0.5039	2.6	1.572	0.7173
-1.0	4.400	0.5127	2.8	1.537	0.7307
-0.8	3.864	0.5218	3.0	1.506	0.7443
-0.6	3.456	0.5312	3.2	1.477	0.7579
-0.4	3.137	0.5408	3.4	1.450	0.7717
-0.2	2.881	0.5508	3.6	1.425	0.7856
0.0	2.671	0.5611	3.8	1.402	0.7996
0.2	2.497	0.5716	4.0	1.381	0.8137
0.4	2.351	0.5825	4.2	1.361	0.8279
0.6	2.226	0.5936	4.4	1.343	0.8422
0.8	2.118	0.6050	4.6	1.325	0.8565
1.0	2.024	0.6166	4.8	1.309	0.8709
1.2	1.943	0.6285	5.0	1.294	0.8854
1.4	1.870	0.6406	5.5	1.260	0.9220
1.6	1.806	0.6539	6.0	1.230	0.9589

(5) If we express the torque L in the following way,

$$L = a_0 + a_1 \cos 2\theta + a_2 \cos 4\theta + a_3 \cos 6\theta + b_1 \sin 2\theta \; b_2 \sin 4\theta + b_3 \sin 6\theta \qquad (30)$$

then the coefficients a_r, b_r are given by

$$a_r = \frac{2}{n} \sum_{o}^{n-1} L_k \cos 2\pi kr/n \qquad (31)$$

$$bb_r = \frac{2}{n} \sum_{o}^{n-1} L_k \sin 2\pi kr/n \qquad (32)$$

where n is the number of torque readings (i.e. if 10° intervals then 18 values are required).

In order to compare the coefficients with theoretical curves or torque expressions, it is usually expressed in terms of sine functions.

i.e. $L = a_0 + A_1 \sin 2(\theta + \alpha_1) + A_2 \sin 4(\theta + \alpha_2)$ (33)

$$\text{where} \quad A_r = (a_r^2 + b_r^2)^{\frac{1}{2}}$$

$$\alpha_r = \tan^{-1} \left(\frac{a_r}{b_r}\right).$$

Let us consider the experimental results in a (110) torque curve of a cubic crystal.

If the results are fitted perfectly by the anisotropy constants K_1, K_2, then the torque $L = A_1 \sin 2\theta + A_2 \sin 4\theta + A_3 \sin 6\theta$ (34)

$$\text{where} \quad A_1 = -K_1/4 - K_2/64$$

$$A_2 = -3K_1/8 - K_2/16$$

$$A_3 = +3K_2/64$$

i.e. if an arbitrary reference point was selected for the torque measurements then $\alpha_1 = \alpha_2 = \alpha_3$ = constant.

The torque curve contains only terms up to 6θ, higher order terms such as 8θ, 10θ are zero.

Typical experimental results for Fe_3O_4 , showing the fit between experimental and theoretical values of coefficients are shown in Table 8.

For the case of torque measurements on crystals which are not cubic, the anisotropy constants are obtained by fitting the Fourier coefficients to the theoretical coefficient appropriate to the symmetry under investigation.

For hexagonal and trigonal crystals, the torque expression for the principal planes are given in Table 9. The Fourier coefficients for torque curves are directly derived from the anisotropy energy expressions given in Table 1.

5. Magnetization curve measurements

As pointed out in the early sections, the anisotropy constants may be determined from the curves of magnetization versus field for single crystals for particular principal crystal directions of magnetizations. The experimental methods used for these measurements are described in detail in Chapter 2. Methods in which the magnetization can be recorded as a function of applied field in different crystal directions in a plane are normally required for anisotropy constant determinations.

Note this method requires the demagnetizing field to be equal in both the easy and hard directions of magnetization.

The curves can be analysed in several ways . The simplest being the initial determination of the easy and hard axes of magnetization. With an instrument such as a vibrating sample magnetometer, in which a direct output is produced that is proportional to the magnetization, the electromagnet providing the magnetizing field B is rotated in the plane under investigation and the directions corresponding to maximum and minimum outputs noted. These directions can be taken as indicating the easy and hard directions respectively. Note these directions usually lie along principal crystal directions, such as <110>,<100> or <111> in cubic crystals. The magnetization curves are then measured along these directions. If a sphere is used as a sample, measurement along the easy direction will produce a linear variation of output with field which suddenly saturates at a field

TABLE 8

Comparison of experimental Fourier coefficients and calculated values using $K_1 = 13.8\ kJm^{-3}$ $(-138 \times 10^3\ erg\ cm^{-3})$ $K_2 = 0$ for the torque curve measured in Fe_3O_4 shown in Figure 11.

	A_1	A_2	A_3
Expt.	$3.45\ kJm^{-3}$	$5.18\ kJm^{-3}$	$-0.1\ kJm^{-3}$
Calc.	$3.45\ kJm^{-3}$	$5.17\ kJm^{-3}$	0

Where the torque amplitudes are expressed per unit volume

TABLE 9

Torque expression for hexagonal and trigonal crystals

Crystal Class		Torque $(-\frac{dE_a}{d\theta})$	Plane
Hexagonal	(1)	$-K_1 \sin 2\theta - K_2 (\sin 2\theta - \frac{1}{2}\sin 4\theta)$ neglecting higher terms	containing C-axis
	(2)	$6K_4 \sin 6\phi$	basal plane
Trigonal	(1)	$-K_1 \sin 2\theta - K_2(\sin 2\theta - \frac{1}{2} \sin 4\theta)$ neglecting higher terms	containing C-axis
	(2)	$6K_3 \sin 6\phi$	basal plane

TABLE 10

Field for saturation along the principal directions in cubic crystals

K_1	B_s along <100>	B_s along <110>	B_s along <111>
+ve	0	$2K_1/M_s$	$\frac{4}{3} M_s(K_1 + \frac{K_2}{3})$
-ve	$-2K_1/M_s$	$-\frac{1}{M_s}(K_1+K_2/2)$	0

greater than the demagnetizing field. The demagnetizing B field is

$$B_D = \mu_o M_s/3 \qquad \left[B_D = \frac{4}{3}\pi M_s\right]_{emu}$$

so that saturating B field is $\mu_0 M_s/3$ thus providing a value for M_s (if the field is accurately known). The field for saturation along the hard direction as defined in Figure 30 will then be expressed in terms of the anisotropy constants, as indicated in Table 10. Thus a simple method to determine the anisotropy constants is to measure the difference in field required to saturate the sample along the easy direction and a principal direction. This can be done easily using the vibrating sample magnetometer in which a voltage proportional to the magnetization is recorded. The applied field can be accurately adjusted to the value just sufficient to saturate the sample in the different directions. From the magnetization curves for a single crystal of magnetite (Fe_3O_4) shown in Figure 31 we can derive the following results. Field for saturation along the <110> direction = 0.0264 T (264 G). Field for saturation along the <100> direction = 0.0524 T (524 G).

Therefore $-\dfrac{2K_1}{M_s} = 0.0524$ T (524 oe)

$$-\frac{K_1}{M_s} + \frac{K_2}{2M_s} = 0.0264 \text{ T} \qquad (264 \text{ oe})$$

From the slope of the magnetization curve, equal to $3/\mu_0$ for a spherical sample ($\frac{3}{4\pi}$ emu) we get

$$M_s = 480 \text{ kAm}^{-1} \text{ (emu)}$$

From which we calculate

$$K_1 = -12.6 \text{ k Jm}^{-3} \quad (12.6 \times 10^4 \text{ erg cm}^{-3})$$

$$K_2 = +0.096 \text{ k Jm}^{-3} \quad (9.6 \times 10^2 \text{ erg cm}^{-3})$$

The anisotropy constants may also be derived from the areas under the magnetization curves. This method was one of the first ways used to derive anisotropy constants.

The area under the magnetization curve is given by

$$\mu_o \int_{o}^{M_s} M \, dH = E_a - E_o \tag{34}$$

where E_o is the free energy in the demagnetized state and E_a is the anisotropy energy in the direction under consideration.

Note in the expressions B is the internal field, that is the applied field corrected for demagnetizing field $B = B_i = B_a - B_d$. Then if the area is measured in the principal directions, i.e. $\langle 100 \rangle$, $\langle 110 \rangle$, $\langle 111 \rangle$ for a cubic crystal.

$$A_{100} = E_{100} - E_o$$

$$A_{110} = E_{110} - E_o$$

$$A_{111} = E_{111} - E_o$$

From the expression for anisotropy energy in equation 16 we get

$$E_{100} = K_o$$

$$E_{110} = K_o + K_1/4$$

$$E_{111} = K_o + K_1/3 + K_2/27$$

Which leads to the following values of K_1 and K_2

$$K_1 = 4(E_{110} - E_{100}) = 4(A_{110} - A_{100}) \tag{35a}$$

$$K_2 = 9(3E_{111} + E_{100} - 4E_{110})$$

$$= 9(-4A_{110} + A_{100} + 3A_{111})$$

$$= 27(A_{111} - A_{100}) - 36(A_{110} - A_{100}) \tag{35b}$$

From the magnetization curves of Fe_3O_4 shown in Figure 31, the values of $A_{110} - A_{100}$, $- A_{111}$ are -3.04, -1.1, k Jm^{-3} (-30.4×10^3, -11×10^3 erg cm^{-3}) respectively giving

$$K = -12.2 \text{ kJm}^{-3} \quad (-12.2 \times 10^4 \text{ erg cm}^{-3})$$

$$K = -2.3 \text{ kJm}^{-3}\text{Jm}^{-3} \quad (-2.3 \times 10^4 \text{ erg cm}^{-3})$$

Note the method has certain disadvantages:
(1) The anisotropy constants are measured over a range of applied fields, and therefore the method is valid only if the constants are independent of applied fields.
(2) Any errors in the magnetization curve are amplified as the constants are given by small differences in large areas.
(3) If the free energy in the demagnetized state is not reproducible and thus not constant for the principal directions, i.e. due to different domain configurations then direct errors will occur.
(4) In cases where the exact form of the anisotropy energy is not known, then the method cannot be applied to determine it, in contrast to the torque method which directly gives the derivative of the energy.

6. Rotating sample magnetometer

Introduction

The Rotating Sample Magnetometer (R.S.M.) outlined in Figure 32 is basically an induction method similar to the Vibrating Sample Magnetometer (V.S.M.) but allows very sensitive measurements of anisotropy to be made with inexpensive equipment. When measuring anisotropy (it can also be used for magnetization measurements), samples are rotated about a symmetry axis in a uniform magnetic field applied normal to the rotation axis. For an anisotropic sample the magnetization will vary in direction around the field direction and a voltage proportional to the magnetization component normal to the magnetic field is sensed by a suitable coil configuration.

The induced voltage V_1 is measured at multiples of the rotation frequency using phase sensitive detection techniques and these produce high sensitivity despite the decrease in output with increase in

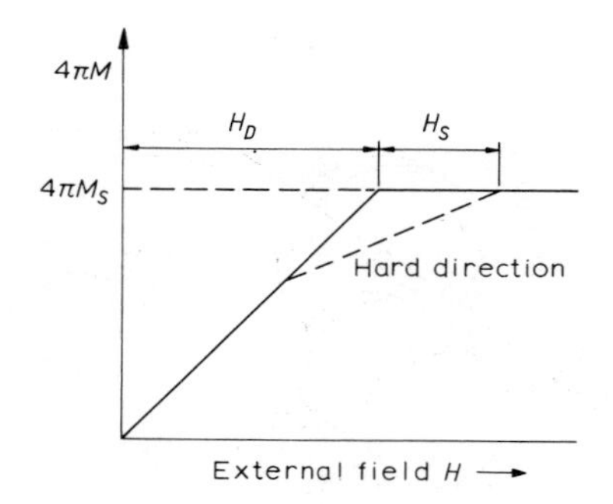

Figure 30 Definition of fields required for saturation in easy and hard directions.

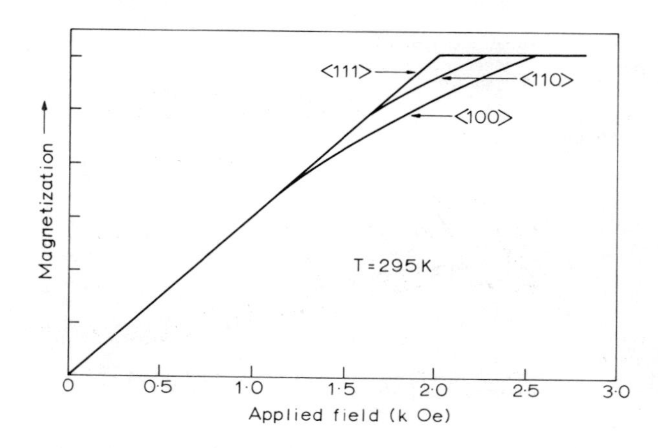

Figure 31 Magnetization curves for a single crystal sphere of magnetite (Fe_3O_4) in the < 111 > , < 110 > and < 100 > directions at room temperature.

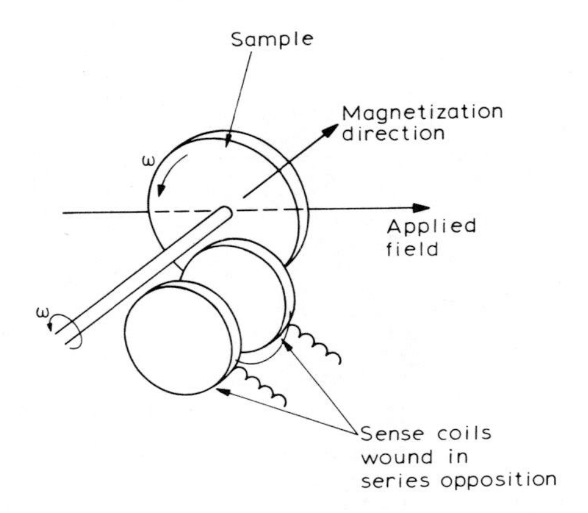

Figure 32 Schematic outline of rotating sample magnetometer.

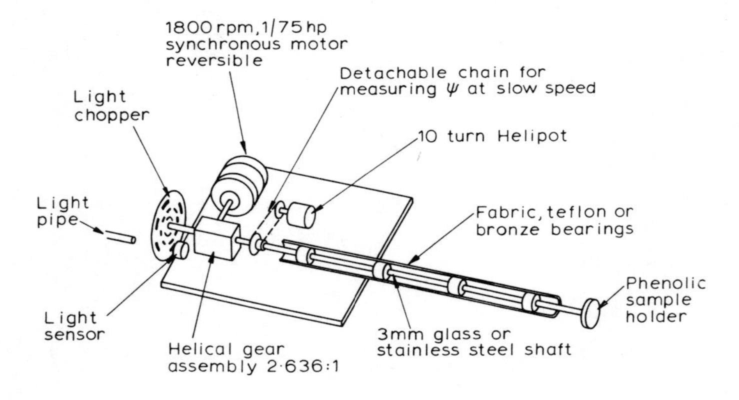

Figure 33. Mechanical details of the rotating sample magnetometer.

(Reproduced by permission of the American Institute of Physics)

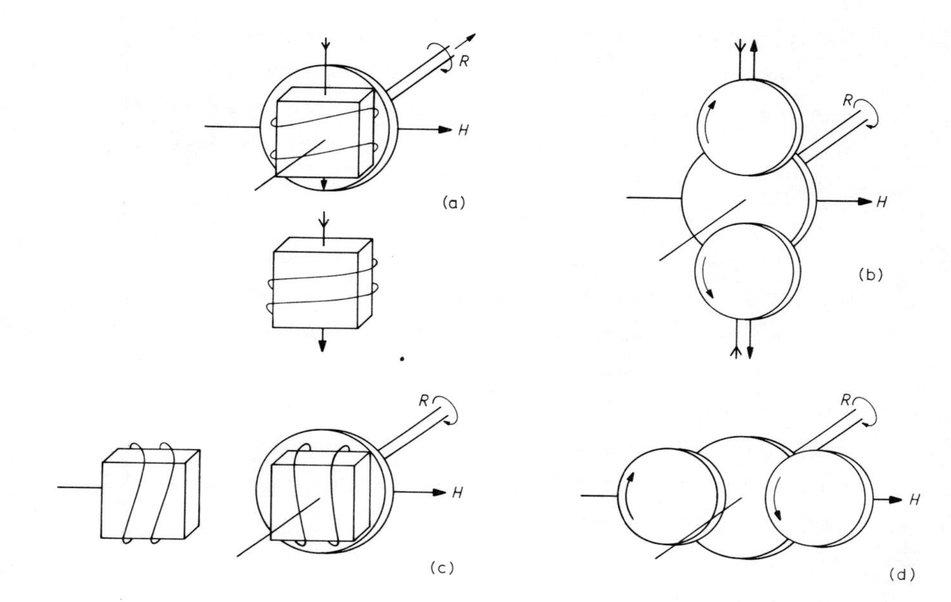

Figure 34 Sample and coil configuration used for anisotropy measurements with the rotating sample magnetometer. Flanders (1970)[24]

(Reproduced by permission of the American Institute of Physics)

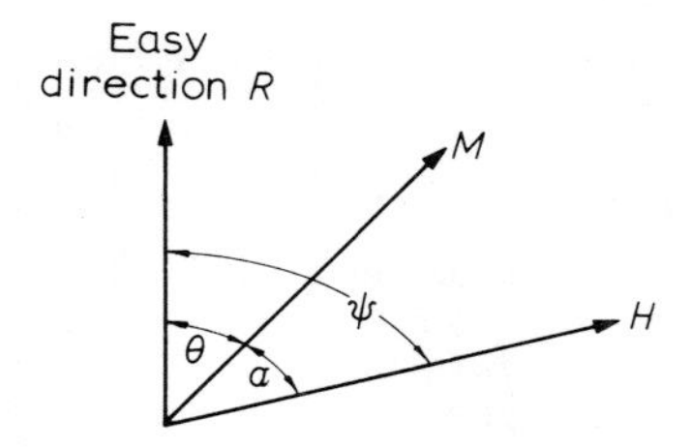

Figure 35 Relative orientation of magnetization, field and easy direction.

applied field. Obviously as B gets very large M and B are nearly always collinear and the output tends to zero. Flanders (1970)[24] has used the instrument extensively and describes how sensitivities of 5×10^{-5} e.m.u. (equivalent to vK/B where B $\geq$ 10 K/M) are easily obtained. Maximum capacity is limited only by the largest sample volume which a rotating system will accommodate; this is to be compared with torque measurements for which the maximum permissible signal (proportional to $K_1 \times v$) is determined by the peak restoring torque, results can be readily converted to give the Fourier components of the anisotropy energy surface.

Note that if the output from the pick up coils is passed into an electronic integrator, the integrated output is proportional to M and has the same shape as the normal torque curve but with its amplitude reduced by a factor B.

Experimental details

Mechanical details of the system are shown in Figure 33. The rotational frequency recommended by Flanders (1970)[24] should be as high as possible yet not coinciding with a harmonic of the supply frequency. He used an 1800 rpm synchronous motor with a step up gear ratio of 2.636:1. For slow rotations (when integration of the output signal is required) the drive is coupled to a multiturn potentiometer to provide a signal proportional to the rotation angle ψ. The reference signal for the phase sensitive detection is provided by a light source, detector and rotating disc shutter. Signals proportional to ω, 2ω, etc. up to 12ω are obtained in this way.

The coils used for detection are mounted in series opposition, and for some cases it is advantageous to be able to rotate them by $\pi/2$ as shown in Figure 34.

Coil shapes that can be used are discs (1 cm dia x 3 mm thick x 18,000 turns (7000 Ω), cylinders 1 cm dia x 1 cm thick x 18,000 turns (7000 Ω), rectangular coils 2.5 cm x 0.8 cm x 2.5 cm long x 10,000 turns 7500 Ω). Flanders states that though the exact number of turns and dimensions are arbitrary, a good choice of orientation, geometry, etc. greatly increases the output. The high coil resistance is matched to the amplifier to the phase sensitive detector with a medium impedance preamplifier.

Admixtures of V and V in the coils can be checked by reversing the direction of rotation. This should just reverse $V_{\parallel}$ and leave $V_{\perp}$ unaltered.

Experimental results

As described above the experimental arrangement measures the harmonic parallel $V_{\parallel}$ and perpendicular $V_{\perp}$ components of the induced voltage due to the rotating components of M_s.

$$\text{Thus } V_{\parallel} = vC \frac{d}{dt} (M_s \cos \alpha) = vC\omega \frac{d}{d\psi} (M_s \cos \alpha) \quad (36a)$$

$$V_{\perp} = vC \frac{d}{dt} (M_s \sin \alpha) = vC\omega \frac{d}{d\psi} (M_s \sin \alpha) \quad (36b)$$

where v is the sample volume

C is a constant depending on the coupling between sample and coil

M_s is the saturation magnetization

ω is the rotational frequency in radians/sec

α, ψ are shown in Figure 35.

From the previous section we derived results for the torque

$$-L = BM_s \sin \alpha = \frac{\partial E_K}{\partial \theta}$$

Integration of the induced voltage V with respect to ψ will give values of output equal to

$$\int V_{\perp} \, dt = vCM_s \sin \alpha = vC \frac{1}{B} \left(\frac{\partial E_K}{\partial \theta}\right)$$

which give the familiar torque curve but with the amplitude reduced by a factor B. However the main feature of the rotating sample magnetometer is that the harmonic components of V can be measured very accurately using conventional phase sensitive detector techniques. These harmonic components can then be directly related to the anisotropy constants of the material. For example let us consider the simple case of a uniaxial material where the anisotropy energy $E_K = K \sin^2\theta$.

$$\text{Then } \frac{\partial E_K}{\partial \theta} = K \sin 2\theta = K \sin 2(\psi - \alpha) = BM_s \sin \alpha$$

This can be expanded as follows, for small values of α, i.e. when $B >> K/M_s$.

$$\frac{\partial E_K}{\partial \theta} = BM_s \sin\alpha = K \sin 2\psi - \frac{K^2}{BM_s} \sin 4\psi$$

The harmonic components of the pick-up voltage are then given by the harmonic components of the expression

$$V = vC\omega\, d/d\psi\, (M\sin\alpha)$$

$$= \frac{vC}{B} \{2K \sin(2\psi + \pi/2) - 4\frac{K^2}{BM_s} \sin(4\psi + \pi/2)\} \tag{37}$$

The 2ω and 4ω components are then given by

$$\begin{aligned} 2\omega &\to vC\omega M\{2(K/BM_s) \sin(2\omega + \pi/2)t\} \\ 4\omega &\to vC\omega M\{-4(K/BM_s)^2 \sin(4\omega + \pi/2)t\} \end{aligned} \tag{38}$$

Flanders[25] has extended this analysis further for both uniaxial and cubic crystals and the results are shown below.

Uniaxial anisotropy - Harmonic components of $V_\perp$

$$\begin{aligned} 2\omega &\to vC\omega M_s\{2(K/BM_s)\} \\ 4\omega &\to vC\omega M_s\{-4(K/BM_s)^2\} \\ 6\omega &\to vC\omega M_s\{+9(K/BM_s)^3\} \\ 8\omega &\to vC\omega M_s\{-\frac{64}{3}(K/BM_s)^4\} \end{aligned} \tag{39}$$

Cubic anisotropy - (110) plane - circular sample

Harmonic components of $V_\perp$

$$2\omega \to -vC\omega M_s[\tfrac{1}{2}\{(K_1/BM_s) + \frac{1}{16}(K_2/BM_s)\}]$$

$$4\omega - -vC\omega M_s\left[3/2\left\{\frac{K_1}{BM_s} + \frac{1}{6}\left(\frac{K_2}{BM_s}\right)^2\right\}\right]$$

$$6\omega - -vC\omega M_s\left[-\frac{27}{16}\left\{\frac{1}{6}\left(\frac{K_2}{BM_s}\right) + \left(\frac{K_1}{BM_s}\right)^2\right\}\right]$$

$$8\omega - -vC\omega M_s\left[-\frac{9}{4}\left\{\left(\frac{K_1}{BM_s}\right)^2 + \frac{2}{3}\left(\frac{K_1}{BM_s}\right)^3\right\}\right]$$

Using this method Flanders (1971)[25] has measured the anisotropy of various materials including Fe_3O_4 and GdIG. For Fe_3O_4, measurements were made in fields greater than $2K/M_s$. $K_1(T)$ is proportional to $V_\perp(2\omega)$, i.e. the 2ω component of the perpendicular component of the induced voltage and its magnitude was obtained from the ratio $V_\perp(8\omega)/V_\perp(2\omega)$ equal to $9K_1/2M_sB$. Results in good agreement with previous torque measurements of Bickford and Syono were obtained. However it should be remembered (as pointed out by Flanders)[24] that despite its versatility, the rotating specimen magnetometer is by no means a substitute for standard measuring instruments. A torque magnetometer is generally more appropriate for obtaining absolute values of anisotropy. Yet the rotating specimen magnetometer is very useful for rapidly surveying material properties and is inexpensive and simple to construct (commercial versions are now available) and despite its high sensitivity it is rugged.

7. Oscillating sample magnetometer

In cases where the anisotropy field of the material to be measured is very large, it may be impossible to saturate the sample in all directions in the crystal. Then it is usually possible to make measurements only near the easy direction of magnetization. This is often the case with uniaxial crystals such as hexagonal garnets and rare earth metals at low temperatures. Here we may not measure the complete torque curves but just the stiffness with which the magnetization is bound to the easy direction.

If we now consider a sample suspended, so that it can perform small oscillations around the direction of the applied field B, then the magnetic contribution to the stiffness is given by

$$c = \frac{d^2F}{d\psi^2}$$

$$= \frac{\partial F}{\partial \psi^2} + 2\left(\frac{\partial^2 F}{\partial\psi\partial\theta}\right) + \left(\frac{\partial^2 F}{\partial\theta^2}\right)\left(\frac{d\theta}{\partial\psi}\right)^2 \quad (41)$$

where ψ , θ are given in Figure 35 and

F the total energy is given in terms of the anisotropy energy by

$$F = E_a - BM_s \cos(\psi - \theta)$$

For oscillations about the easy direction we can put $\psi = 0$ and therefore we get for a disc shaped sample of uniaxial material in which $E_a = K \sin^2\theta$

$$\text{Stiffness } c = \left(\frac{1}{M_sB} + \frac{1}{2K}\right)^{-1} \quad (42a)$$

This can be also expressed in terms of the anisotropy field H_A

$$c = M_s \left(\frac{1}{B} + \frac{1}{B_a}\right)^{-1} \quad (42b)$$

Note for cubic crystals, if $\langle 100 \rangle$ is the easy direction

$$B_a = 2K_1/M_s$$

and if $\langle 111 \rangle$ is easy

$$B_a = -\frac{4}{9}\left(\frac{3K_1 + K_2}{M_s}\right)$$

In practice the sample is mounted in a torsional pendulum whose frequency is determined in the zero field and with the external field applied.

The change in the reciprocal period of oscillation squared (i.e. $\Delta \frac{1}{T^2}$) is then given in terms of the magnetic stiffness and the moment of inertia I by

$$\Delta\left(\frac{1}{T^2}\right) = \frac{c}{4\pi^2 I} \tag{43}$$

Using the above expression for c we see that a plot of c^{-1} against $\left(\frac{1}{B}\right)$ give a straight line with slope $\left(\frac{1}{M_s}\right)$ and intercept $\left(\frac{1}{B_A}\right)$. This method was first used by Rathenau and Snoek (1941)[26] and has since widely used to determine anisotropy constants of hexagonal ferrites. A modification, in which the torsional pendulum was self excited at a high frequency (100 c/s) for measurements at elevated temperatures has been described by Zijlstra (1961)[27].

8. Microwave resonance measurements

In a previous section (2.2) it was pointed out that anisotropy constant may be derived from microwave resonance measurements on single crystals.

The condition for resonance can be expressed in the form

$$\omega = \gamma B_{eff} \qquad \gamma = \frac{ge}{2mc} \tag{44}$$

where ω is the frequency of the oscillating magnetic field which excites the resonance, and γ is given in terms of the g-factor and the usual constants.

B_{eff} contains contributions from the applied field as well as terms which depend on the anisotropy of the specimen.

For cubic crystals, the value of the anisotropy constants can be determined from measurements of the applied field required to produce resonance at a particular frequency. Thus for the <100> , <110>and <111> directions the resonance conditions for a sphere are:

$$\langle 100\rangle, \quad B^2_{eff} = (B_{100} + 2K_1/M_s)^2 \tag{45a}$$

$$\langle 110\rangle, \quad B^2_{eff} = \left(B_{110} - \frac{2K_1}{M_s}\right)\left(B_{110} + \frac{K_1}{M_s} + \frac{K_2}{2M_s}\right) \tag{45b}$$

$$\langle 111\rangle, \quad B^2_{eff} = \left(B_{111} - \frac{4}{3}\frac{K_1}{M_s} - \frac{4}{9}\frac{K_2}{M_s}\right)^2 \tag{45c}$$

Knowing the values of B_{100}, B_{110}, B_{111} then $\frac{K_1}{M_s}$ and $\frac{K_2}{M_s}$ can be derived. To derive K_1 or K_2 the value of M_s must also be known. If the g values are not assumed to be isotropic, then measurements must be made at two microwave frequencies to determine the anisotropy constants. For hexagonal and trigonal crystals in the form of spheres the resonance expressions are given as follows.

Hexagonal crystals

Field applied parallel to c-axis.

$$B_{eff} = (B + B_a) \qquad (46a)$$

Field applied perpendicular to c-axis, i.e. in the basal plane.

(a) $B < B_a$

$$B^2_{eff} = (B_a^{\,2} - B^2) \qquad (46b)$$

(b) $B < B_a$

$$B^2_{eff} = B(B - B_a)$$

$$B_a = \frac{2K_1}{M} \qquad (46c)$$

Trigonal crystals

Field applied in basal plane (assumed to contain the easy direction)

$$B^2_{eff} = B(B + B_a) \qquad (47a)$$

neglecting anisotropy in basal plane $B_a = -\frac{2K_1}{M}$

Incorporating higher order terms.
Then for a spherical sample we get

$$B^2_{eff} = (B + B_a - \frac{B^A}{6} \cos 6\beta)(B - B^A \cos 6\beta) \qquad (47b)$$

where

$$B_a = -(2K_1 + 4K_2 + 6K_3')/M$$

$$B^A = 36\ K_3/M$$

The accuracy of the method depends on the linewidth of the ferromagnetic resonance in the sample that is being measured, and this method has been more useful in measuring anisotropy in low linewidth materials such as the ferromagnetic garnets than in metallic materials which possess much broader linewidths.

The method is very sensitive to higher order terms in the anisotropy, as it depends on the second derivative of the anisotropy energy, and in cases where a check on the validity of the anisotropy energy expression is required, measurements must be made not only at principal directions but as a function of angle in principal planes. If this is required, the resonance curves must be corrected for the difference in direction of the applied field and the magnetization vector before attempting to fit anisotropy constants. Although this correction can be derived from resonance results alone by an iterative process, the results are not very accurate, and the procedure is made easier by using data obtained independently from torque measurements.

Finally the results obtained from resonance measurements need not necessarily agree with the values of anisotropy constants determined by other methods at very low frequencies. Thus where relaxation effects are present in the material the anisotropy is time dependent and will depend on the frequency at which the measurements are made. These effects are now considered in more detail.

9. Measurement of induced anisotropy and relaxation effects

In many magnetic materials a uniaxial magnetic anisotropy may be created by annealing the sample at a high temperature in a magnetic field large enough to cause saturation. This process is usually referred to as a magnetic anneal. This induced anisotropy is superimposed on the intrinsic magnetic crystalline anisotropy energy of the material, and has associated with it a relaxation time.

It may be shown from symmetry considerations that the form of the induced uniaxial anisotropy energy in cubic crystals is given by

$$E_a = -F(\alpha_1^2\beta_1^2 + \alpha_2^2\beta_2^2 + \alpha_3^2\beta_3^2)$$
$$-G(\alpha_1\alpha_2\beta_1\beta_2 + \alpha_2\alpha_3\beta_2\beta_3 + \alpha_3\alpha_1\beta_3\beta_1) \tag{48}$$

where $\beta_1, \beta_2\ \beta_3$ are direction cosines of magnetization during annealing and $\alpha_1, \alpha_2, \alpha_3$, are direction cosines of the magnetization during measurements. As the free energy will always decrease on annealing, the constants F and G are always positive.

These constants F, G play an important part in the study of induced anisotropy and magnetic after effect. Their significance is discussed in reviews by Slonczewski (1963)[28] and Braginski (1965)[29]

To evaluate F and G experimentally, the most obvious method is to use the torque magnetometer.

The torque in the principal planes is given in Table 11 in terms of the measurement angle (θ) and annealing angle ψ in those planes.

Then if the sample is annealed at a high temperature and quenched to a lower temperature where the induced anisotropy has a long relaxation time, a conventional torque curve can be measured at this temperature. Using the expressions in Table 11 the values of F and G can be derived. Or more simply we can then see that if measurements are carried out in the (100) plane that F is given directly by the torque measured in the $\langle 110 \rangle$ direction after annealing in the $\langle 100 \rangle$ direction. Similarly G is given by the torque in the $\langle 100 \rangle$ direction after annealing in the $\langle 110 \rangle$ direction.

The induced anisotropy coefficients F and G may also be deduced from microwave resonance measurements. General results of the resonance fields corresponding to torque curves in Table 11 are given in the paper by Teale (1967)[30] Some special results are given in Table 12. However Teale has also considered the case of shorter relaxation times and the cases where these can cause a shift in the effective field for resonance and a contribution to the linewidth ΔH. In a similar way, Broese van Groenou, Page and Pearson (1969)[31] have also calculated the effect in torque curves of a time dependent anisotropy. If we consider a torque experiment in which the magnet

TABLE 11

Torque in principal plane in cubic crystals due to magnetic anneal

(100)

$$L = -F \cos 2\psi \sin 2\theta + \frac{G}{2} \sin 2\psi \cos 2\theta - \frac{K_1}{2} \sin 4\theta$$

(110)

$$L = -\left\{\frac{K_1}{4} - \frac{K_2}{64} - \frac{F}{4} - \frac{3F}{4} \cos 2\psi + \frac{G}{8} (1 - \cos 2\psi)\right\} \sin 2\theta$$

$$+ \frac{G}{2} \sin 2\psi \cos 2\theta$$

$$- \left(\frac{3K_1}{8} - \frac{K_2}{16}\right) \sin 4\theta + \frac{3K_2}{64} \sin 6\theta$$

θ is direction of magnetization during measurement

ψ is direction of magnetization during anneal.

Angles measured from [100] .

TABLE 12

Resonance conditions in cubic crystals due to magnetic anneal

Anneal direction	Measurement direction	Resonance Condition
[100]	[100]	$\frac{\omega}{\gamma} = B + \frac{2K}{M_s} + \frac{2F}{M_s}$
[111]	[100]	$\frac{\omega}{\gamma} = B + \frac{2K}{M_s}$
[111]	[111]	$\frac{\omega}{\gamma} = B - \frac{4K}{3M_s} + \frac{G}{M_s}$
[111]	[111]	$\frac{\omega}{\gamma} = B - \frac{4}{3}\frac{K}{M_s} - \frac{G}{3M_s}$

where B is the field required for resonance, K is the anisotropy coefficient corresponding to the intrinsic cubic contribution and normally referred to as K_1

rotates with angular velocity ω while the anisotropy is time dependent with a relaxation time τ, then in the case when $\omega\tau >> 1$ and the experiment is done at a time $t < \tau$ the theory predicts a rotational hysteresis of magnitude.

$$\Delta T = \{C_2 + C_3 + (C_3 - C_2) \cos 4\phi\} \left(\frac{4\omega\tau}{1 + 4\omega^2\tau^2}\right) \tag{49a}$$

where the coefficients C_i depend on the plane under consideration and are related to F and G as shown in Table 13, and an additional torque

$$T_{av}(\phi) = 2C_1 \sin 2\phi + \frac{(C_2 - C_3) \sin 4\phi}{(1 + 4\omega^2\tau^2)} \tag{49b}$$

which could be confused with an intrinsic K_1 or K_2 term. However in all this analysis, one must not stick too blindly to the phenomenological treatments; and if a microscopic treatment can be found, this should be compared directly with torque curves and resonance measurements, rather than trying to fit the results to anisotropy coefficients that have no physical significance.

For frequencies between d.c. and microwaves, other methods of measurement are to be preferred. An apparatus has been described which will measure anisotropy fields at frequencies in the range 2kHz to 50kHz in small single crystal spheres or discs. In the method, shown in Figure 36, the sample is arranged so that it can be rotated about a vertical <110> axis and a saturating field is applied in the horizontal (110) plane. A small low frequency alternating field is applied parallel to the rotation axis, this caused the magnetization to oscillate through a small angle in a plane normal to the (110) plane the magnitude of which was sensed by a secondary coil as an alternating voltage v. Then for spherical samples

$$v - v_o = A/(B + B_a) \tag{50}$$

where v_o, A are instrumental constants, B is the applied field and B_a is the anisotropy field.

Where relaxation effects occur, B_a may vary with frequency. The feature of the method is that the exciting frequency is always so low (<50kHz) that the magnetization in making small oscillating

TABLE 13

Expressions for induced anisotropy in different planes for cubic crystals

$$E_u = C_o + C_1 (\cos 2\phi + \cos 2\psi) + C_2 \cos 2\phi \cos 2\psi + C_3 \sin 2\phi \sin 2\psi$$

where ϕ and ψ are angles made by magnetization and annealing field with reference direction.

(001) ϕ and ψ measured from [100]

$C_o = C_2 = -F/2$ $\quad$ $C_1 = 0$ $\quad$ $C_3 = -G/4$

(011) ϕ and ψ measured from [100]

$C_o = C_2 = -3F/8 - G/16$

$C_1 = -F/8 + G/16$ $\quad$ $C_3 = -G/4$

(111) ϕ and ψ are measured from $[1\bar{1}0]$ axis

$C_o = -F/3 - G/12$ $\quad$ $C_1 = 0$

$C_2 = C_3 = -(F + G)/6$

For a polycrystalline sample

$E_u = -K_u \cos^2 (\phi - \psi)$

$C_o = C_2 = C_3 = -Ku/2$ $\quad$ $C_1 = 0$

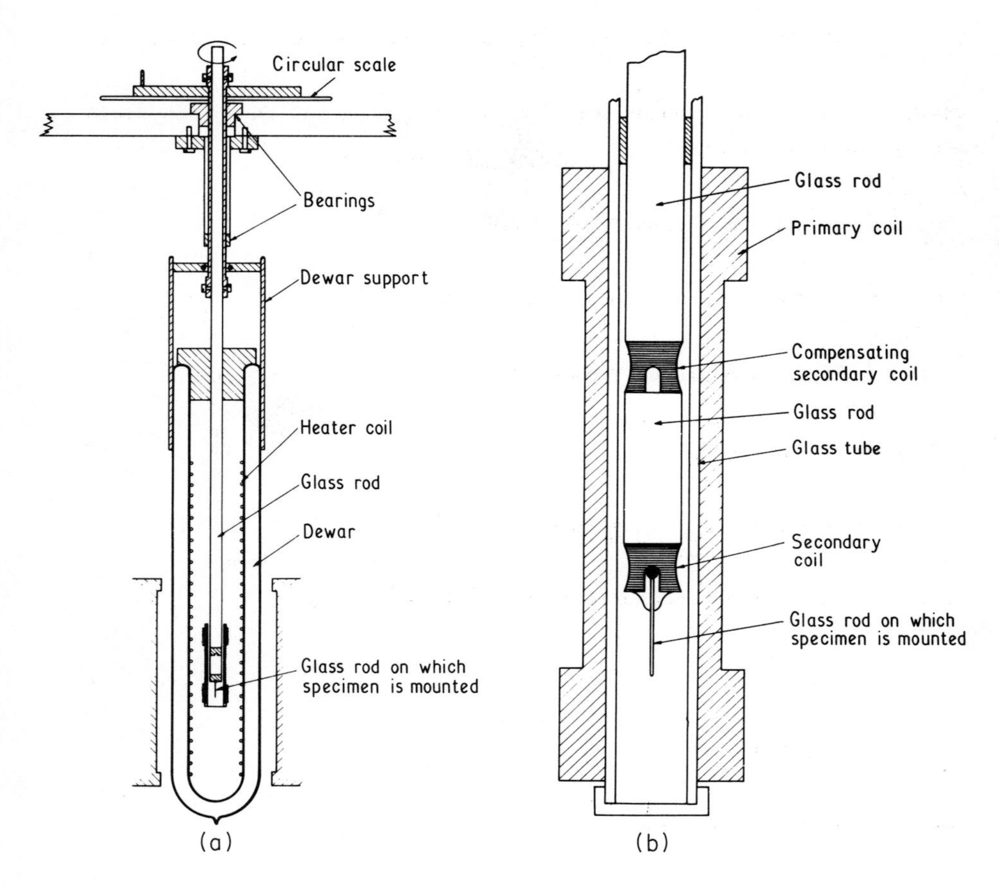

Figure 36 Measurement of induced anisotropy due to Knowles (1969)[32]
(a) general arrangement of coils, magnet pole pieces and Dewar,
(b) arrangement of primary and secondary coils.

(Reproduced by permission of the Institute of Physics)

deviations from a principal plane (usually (110) always lies in a plane perpendicular to the plane whereas at microwave frequencies it will precess about an axis in the plane.

The experimental method consists of setting the applied field B parallel to [100] its value is then determined i.e. B_{100} corresponding to the voltage ouptut v_{100}. The sample is then rotated so that B is parallel to [110] and B_{110} adjusted until $v_{110} = v_{100}$. Then if $B_{100} - B_{110}$ is plotted as a function of frequency, the course of the relaxation may be examined and estimates of the induced anisotropies and time constants may be made.

Using the method Knowles [32] has measured the effect on K_1 of relaxation in Si doped yttrium iron garnet (YIG) at temperatures down to 77K.

Previously Knowles [33] had also measured small changes in B_a due to magnetic anneal effects in a single crystal of Fe_3O_4. Induced anisotropies of 4J m^{-3}(40 ergs cm^{-3}) compared to intrinsic values of $K_1 \sim (10^4 J\ m^{-3}$ (10^5 ergs cm^{-3}) were measured demonstrating the high sensitivity of the method.

A possible improvement on this method using a mechanical transducer to oscillate the magnetization is the plane of the applied field has been described recently by V. Frank and O.V. Nielsen [34]. A schematic diagram of the apparatus is shown in Figure 37. In this method the induced voltage is proportional to $B_a/B_a + B)$. However care must be taken in the detector coil construction and mounting to avoid parasitic signals. This method has been used to measure anisotropy in YIG and Sn doped haematite.

10. Effect of stress on anisotropy measurements

All the analysis of results we have considered so far has considered the case of a rigid crystal, the anisotropy constants being thus defined for the case of constant strain. These values would only be measured in practice if the crystal was clamped such that its lattice dimension remained constant. If the lattice is allowed to deform during measurement the free energy will contain additional terms which depend on the strain in the crystal. In fact the crystal will deform spontaneously if by so doing the

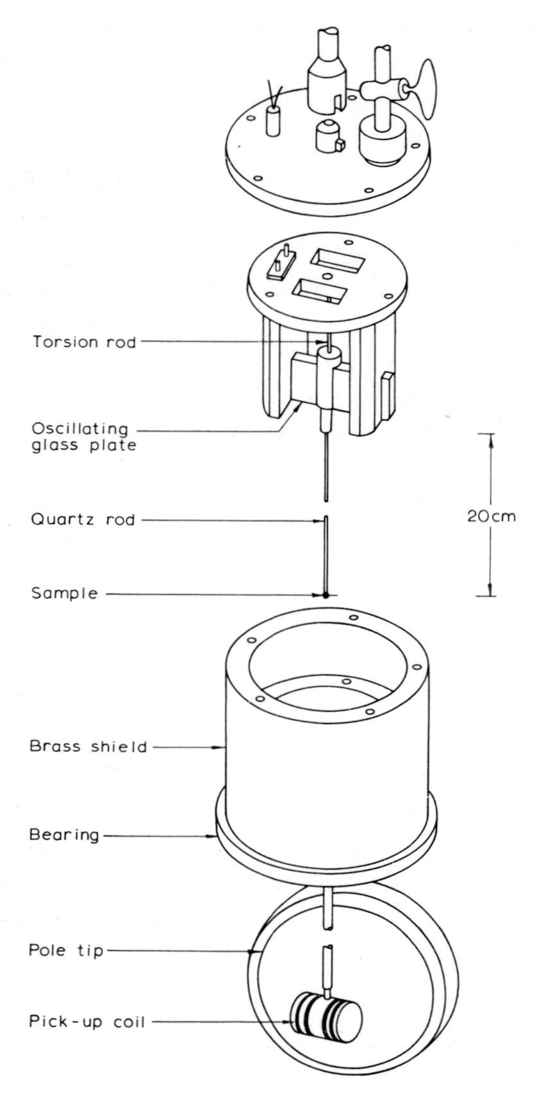

Figure 37 Exploded view of oscillating sample anisotropy meter for room temperature measurements. Frank and Nielson (1971)[34]

(Reproduced by permission of the Institute of Physics)

anisotropy energy is reduced. This deformation is the origin of the phenomena known as magnetostriction and is thus the reason why any theoretical treatment of anisotropy can also explain the occurrence of magnetostriction.

In this section we shall show how the anisotropy energies at constant stress and constant strain are related. The total free energy of the crystal is made up of

(1) The crystal energy E_K

(2) The magnetostriction energy E_M

(3) The elastic energy E_L,

which depend on the direction of the magnetization α_i and the components of the strain tensor A_{ij} as follows

$$E_K = K_1(\alpha_1^2\alpha_2^2 + \alpha_2^2\alpha_3^2 + \alpha_3^2\alpha_1^2) \tag{51a}$$

$$E_M = b_{ij}A_{ij} + b_{ijk}\alpha_i A_{jk} + b_{ijkl}\alpha_1\alpha_2 A_{kl} \tag{51b}$$

$$E_L = c_{ijkl} A_{ij} A_{kl} \tag{51c}$$

Considering the case of cubic crystals and applying the appropriate symmetry operations

$$E_M = b_{1111}(\alpha_1^2 A_{11} + \alpha_2^2 A_{22} + \alpha_3^2 A_{33}) + b_{1212}(\alpha_1\alpha_2 A_{12} + \alpha_2\alpha_3 A_{23} + \alpha_3\alpha_1 A_{31}) \tag{52a}$$

$$E_L = \tfrac{1}{2}c_{11}(A_{11}^2 + A_{22}^2 + A_{33}^2) + c_{12}(A_{11}A_{22} + A_{22}A_{33} + A_{33}A_{11}) + 2c_{44}(A_{12}^2 + A_{13}^2 + A_{23})^2 \tag{52b}$$

For equilibrium we set $\dfrac{\partial E}{\partial A_{ij}} = 0$

This gives six equations.

$$1. \quad \partial E/\partial A_{11} = c_{11}A_{11} + c_{12}(A_{22} + A_{33}) + b_{1111}\alpha_1^{\;2} = 0$$

$$2. \quad \partial E/\partial A_{22} = c_{11}A_{22} + c_{12}(A_{33} + A_{11}) + b_{1111}\alpha_2^{\;2} = 0$$

$$3. \quad \partial E/\partial A_{33} = c_{11}A_{33} + c_{12}(A_{11} + A_{33}) + b_{1111}\alpha_3^{\;2} = 0$$

$$4. \quad \partial E/\partial A_{12} = 4c_{44}A_{12} + b_{1212}\alpha_1\alpha_2$$

$$5. \quad \partial E/\partial A_{13} = 4c_{44}A_{13} + b_{1212}\alpha_1\alpha_3$$

$$6. \quad \partial E/\partial A_{23} = 4c_{44}A_{23} + b_{1212}\alpha_1\alpha_3$$

Giving as solutions

$$A_{11} = \frac{b_{111}\{c_{12} - (c_{11} + 2c_{12})\alpha_1^{\;2}\}}{(c_{11} - c_{12})(c_{11} + 2c_{12})}$$

$$A_{ij} = \frac{-b_{1212}\,\alpha_i\,\alpha_j}{4c_{44}}$$

Inserting these values back into the expression for the energy we get

$$E = (K + \Delta K)(\alpha_1^{\;2}\alpha_2^{\;2} + \alpha_2^{\;2}\alpha_3^{\;2} + \alpha_1^{\;2}\alpha_1^{\;2}) \qquad (53)$$

where ΔK is the difference in anisotropy at constant stress and constant strain

$$\Delta K = \frac{9}{4}((c_{11} - c_{22})\lambda^2_{100} - 2c_{44}\lambda_{111}^{\;2}) \qquad (54)$$

where $\lambda_{100} = - \frac{2}{3} B_1/(c_{11} - c_{22})$

$\lambda_{111} = - \frac{1}{3} B_2/c_{44}$

and B_1, B_2, the usual magneto elastic constants are equal to

b_{111} and $\frac{b_{1212}}{2}$ respectively.

If terms in K_2 are required then a higher order term analysis by Baltzer (1957)[35] gives

$$\Delta K_1 = (c_{11} - c_{12})(h_1^2 + \frac{7}{3} h_1 h_4 - h_1 h_6 - h_4 h_6)$$

$$- 2c_{44}h_2^2 + 3(c_{11} + c_{12})(h_3^2/5 + h_3 h_8/105) \qquad (55)$$

$$\Delta K_2 = -3(c_{11} - c_{12})h_1 h_4$$

$$+ 12c_{44}h_2 h_5 + 3(c_{11} + c_{12})(h_3 h_8/5 + h_8^2/105) \qquad (56)$$

where strain is expressed in terms of 9 magnetostriction coefficients h_1 to h_9.

An example of the corrections involved can be found using results of magnetostriction and elastic constants of some well known materials.

Iron	(Fe)	$\Delta K_1/K_1$	=	2×10^{-3}
Nickel	(Ni)	$\Delta K_1/K_1$	=	10^{-2}
Magnetite	(Fe_3O_4)	$\Delta K_1/K_1$	=	0.3
Cobalt substituted manganese ferrite	$Co_{.25}Mn_{.75}Fe_2O_4$	$\frac{\Delta K_1}{K_1}$	=	0.3

Obviously for materials with large magnetostriction, the stress contribution to the anisotropy can be quite large. Similar calculations giving the results of the free energy at constant strain and constant stress for hexagonal, tetragonal and orthorhombic crystals are given by Mason (1954)[36]

11. Sample preparation

The method of preparation of samples is of great importance in measurements of anisotropy. Usually the sample is required to be single crystal and if the crystal anisotropy is small, effects of strains introduced during preparation should be kept at a minimum. The methods used in practice differ according to the material under investigation.

The range of single crystals of magnetic materials that are available or have been grown is very great and fortunately there exists a bibliography of available samples. This is produced by the Oak Ridge Laboratory and given in Table 14 which also gives a summary of crystal growing methods used to grow some well known magnetic materials.

Methods of preparation of samples from single crystals involve fairly standard metallurgical techniques. However for many measurements a spherical sample is desirable, and the details of how a sphere of size 0.3 mm - 8 mm is given as follows. The manufacture of a polished sphere requires three or four stages. Firstly the selection and initial shaping of the material, secondly producing a rough sphere by blowing in an abrasive lined container, thirdly grinding and polishing using a machine with two rotating tubes and lastly if necessary, polishing between two rotating pads. The following details have been applied to different materials with slight modification.

(1) Selection and initial preparation of material

Ideally the material for a sphere should take the form of a cube cut from good, sound material. When using flux grown materials, this ideal shape may not be possible. Sometimes it may be necessary to use the whole of a small crystal if (a) all the crystals are small or (b) variations in chemical composition occur between small and large crystals. At this stage surface defects and irregularities can be removed by careful hand grinding with an abrasive powder.

(2) Rough preparation

The next step is to achieve a rough sphere with variations in diameter of not more than 1 in 25. This is obtained by blowing the material round inside an abrasive lined vessel (the 'blower') using compressed air. Spheres of diameter less than 4 mm can be shaped in an emergy paper lined blower shown in Figure 38 with a cavity 1¼ diameter and ½" high similar to that described by Bond (1951)[37].

The grade of emergy paper used depends on the size of the sphere, shape of crystal and finish required. For spheres over 4 mm diameter a

TABLE 14

showing crystal growing methods used to grow some well known magnetic materials

		Method
Fe	-	strain anneal
Ni	-	Bridgeman/stationary solidification
Co	-	Bridgeman
Fe_3O_4	-	Bridgeman
YIG ($Y_3Fe_5O_{12}$) garnet	-	Flux
$MnFe_2O_4$	-	Bridgeman
$NiFe_2O_4$	-	Flux
Gd	-	Bridgeman
$YFeO_3$ (ortho-ferrite)	-	Flux
$BaFe_{12}O_{19}$	-	Flux

Information on crystal suppliers or reference on crystal growth and preparation can be obtained on request from the Research Materials Information Centre, Oak Ridge National Laboratory, Oak Ridge, Tennessee, 37830, U.S.A.
Further information may be obtained from books by Laudise[43] Pamplin[44].

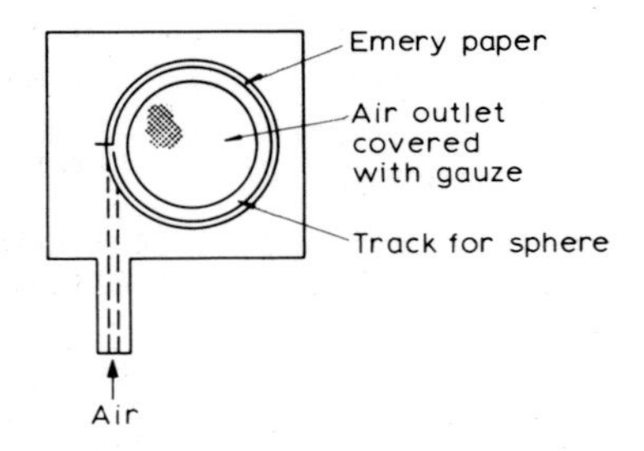

Figure 38 Emery paper sphere blower.

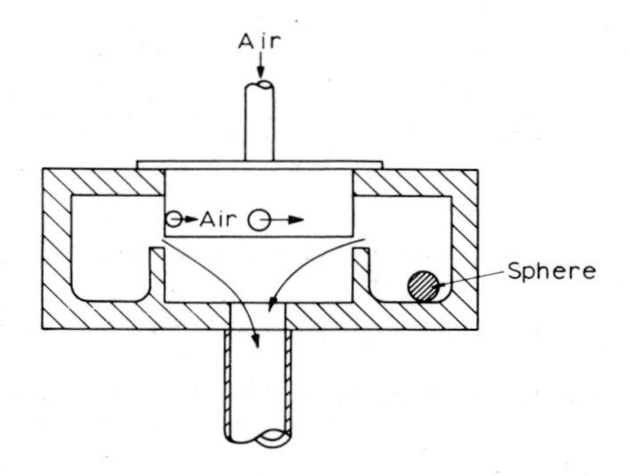

Figure 39 Carborundum loaded araldite sphere blower.

larger blower is required. This is to prevent the material breaking up due to the larger forces encountered in a smaller blower. In this case the blower shown in Figure 39 can be used. Here the track is made from carborundum loaded araldite. When sufficient sphericity has been achieved it is then possible to move on to the two tubes machine.

(3) Two tubes grinding and polishing

In the two tubes polisher described by Cross (1961)[38] the sphere is held between a rotating rubber rimmed wheel and two centre rotating nickel tubes which subtend an angle of about 135° to each other. The sphere thus rotates in a random manner and when coated with abrasive it will be ground away on the high spots only. This is shown more clearly in Figure 40 and Figure 41. When starting to grind a sphere tube of diameter about that of the sphere should be selected, and care should be taken to keep the ends of the tube from becoming rounded by the abrasive. Diamond paste is usually used as the abrasive and grades from 6 micron down to 1 micron are employed in preparation of spheres of garnets and ferrites.

For extra fine polishing, a polisher similar to that described by Lam (1965)[39] are shown in Figure 42.

The sphere is sandwiched between the ends of two vertical rods fixed with polishing pads and one of the rods is rotated at about 1000 rpm (1/10 micron) diamond paste is used) very good polishing can be achieved.

Using these techniques, spheres can be made to sphericities better than 1 in 10,000. The methods have been also successful with materials possessing a large anisotropy of hardness such as the hexagonal ferrites.

(4) Finally the crystal sphere usually needs to be mounted in the apparatus in a particular crystallographic orientation. After transfer from a suitable method of crystal orientation (i.e. X-ray diffractometer) the sphere is cemented on to the sample holder. Some details of adhesives which have been successfully used for this purpose are as follows.

Adhesives for sphere mounting

Araldite - epoxy resin supplied by CIBA

Resin type AV 121.

Hardener type HY956.

Cured cold for 16 hours and then at 70°C for 30 minutes. This adhesive undergoes a very small change in dimensions on hardening, is non magnetic, does not etch the samples and has been found to be satisfactory for mounting of garnet crystals for magnetic experiments over a wide range of temperature from 1.5°K upwards.

12. Methods suitable for polycrystalline materials

When it is difficult to obtain single crystals of a magnetic material, it is still possible to obtain information on the anisotropy constants of the material. Thus for a material possessing a random arrangement of crystallites, certain parts of the magnetization curve such as approach to saturation, resonance, and initial permeability are very dependent on the anisotropy of the crystallites. However these methods depend on assumptions regarding the magnetization processes, and these usually involve only reversible rotation of the magnetization vector. For this reason, the method using the approach to saturation, at fields well above the anisotropy field seems to give the most reliable results. Thus if the field is sufficiently high that no domain walls are present, the magnetization can be expressed in terms of the internal field H by an expression of the form

$$M = M_s \left(1 - \frac{a}{H^2} - \frac{b}{H^3}\right) \tag{57}$$

where for cubic crystals with <111> easy directions, Becker and Doring (1939) show that

$$a = 0.07619\left(\frac{K_1}{\mu_o M_s}\right)^2 \qquad \left[a = 0.07619\left(\frac{K_1}{M_s}\right)^2\right]_{emu}$$

$$b = 0.0384\left(\frac{K_1}{\mu_o M_s}\right)^3 \qquad \left[b = 0.0384\left(\frac{K_1}{M_s}\right)^3\right]_{emu}$$

where K_1 is the first order anisotropy constant, M_s is the saturation magnetization and H the internal field acting on the material. The expression is valid in relatively high fields. However it neglects the effect of interaction between the crystallites and thus gives an underestimate of K_1. The coefficients a and b can be derived from careful measurements of the magnetization, and Figure 43 shows results of approach to saturation for a manganese zinc ferrite given by Chamberlain (1961)[40]. In this figure $(M_s - M)H^3$ is plotted as a

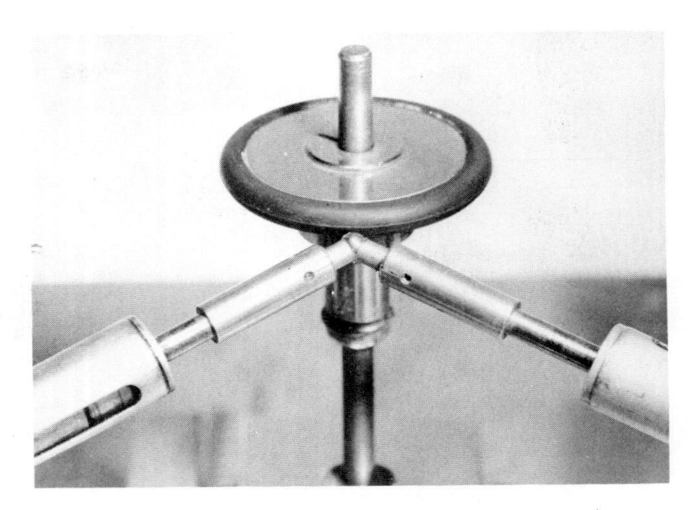

Figure 40 Location of sphere in polishing machine.

(Reproduced by permission of the American Institute of Physics)

Figure 41 General arrangement of sphere polishing machine.

(Reproduced by permission of the American Institute of Physics)

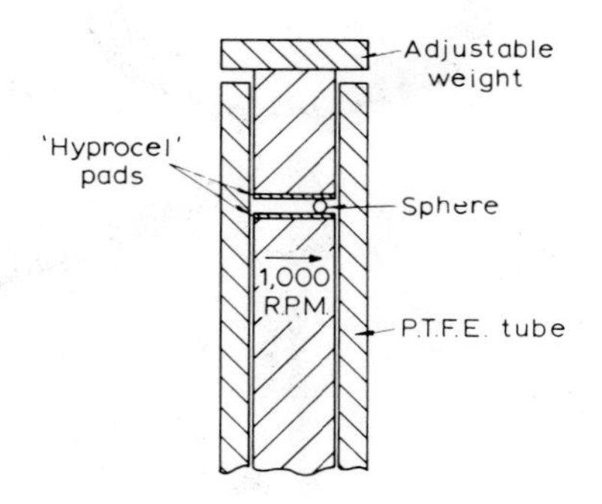

Figure 42 Rotating rod polisher used for final finishing of spheres.

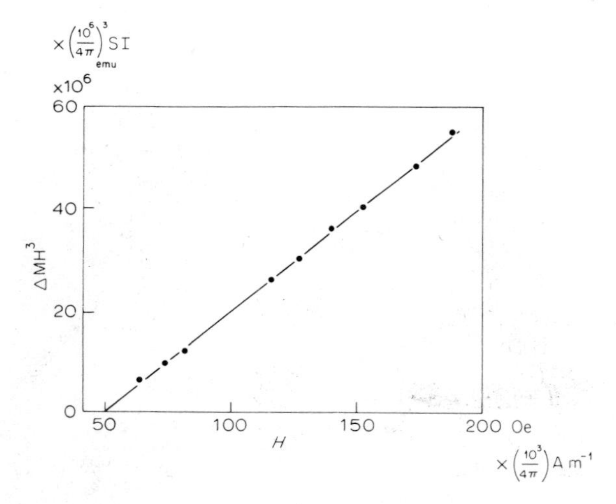

Figure 43 Results of $(M_s - M)H^3$ plotted against H derived from approach to saturation for manganese zinc ferrite (Chamberlain 1961)[40]

function of H. The slope of the straight line gives the coefficient a and the intercept gives a value for b. The negative value of b indicates that K_1 is negative and that the easy direction of magnetization is <111>. Similar derivations of anisotropy constants from measurements of the approach to saturation of polycrystalline ferrites have been made by Dionne (1969)[41].

Initial susceptibility

If the initial susceptibility in a polycrystalline material is considered to be due only to rotations of the magnetization vector towards the applied field, then a simple expression for the magnetic susceptibility can be deduced in terms of K_1 and M_s for the material, e.g.

$$\chi^{rot} = \frac{\mu_o M_s^2}{2K_1} \qquad \left[\chi^{rot} = \frac{M_s^2}{2K_1}\right]_{emu} \tag{58}$$

At high frequencies , the effective anisotropy field will give rise to a resonance frequency.

$$\omega^{rot} = \gamma B_{anis} = \gamma \frac{4}{3} \frac{K_1}{M_s} \tag{59a}$$

$$f^{rot} = \gamma \frac{2}{3\pi} \frac{K_1}{M_s} \tag{59b}$$

We can thus use the frequency spectrum of the susceptibility to deduce K_1. This approach has been used by Cunningham (1963)[42] to deduce the anisotropy of scandium substituted garnets from measurements on polycrystalline samples.

REFERENCES

1. R.R. Birss "Symmetry and Magnetism", North Holland Publishing Company, Amsterdam 1964, pp 158 - 162
2. M.I. Darby and E.D. Isaac, I.E.E.E. Trans on Magn., Vol MAG 10 259 (1974)
3. F.W. Harrison, J. Sci. Instr., 33, 5 (1956)
4. R.F. Penoyer, Rev. Sci. Instr., 30, 711 (1959)
5. A.A. Aldenkamp, C.P. Marks and H. Zijlstra Rev. Sci. Instr., 31, 544 (1960)
6. G. Aubert, J. App. Phys., 39, 504 (1968)
7. P. Escudier, Ann de Physique 9 125 (1975)
8. R.R. Birss and P.M. Wallis J. Sci. Instr., 40, 551 (1963)
9. R.R. Birss and B.C. Hegarty J. Sci. Instr., 44, 621 (1967)
10. R. Gerber and F. Vilim, J. Sci. Instr, Series 2, vol 1, 389 (1968)
11. E.J. Fletcher, A. de Sa, W. O'Reilly and S.K. Banerjee, J.Sci. Instr., Series 2, Vol 2, 311 (1969)
12. R.F. Pearson, J. Phys. Rad., 20, 409 (1959)
13. J. Bransky, A.A. Hirsch, I. Bransky, J. Sci. Instr., Series 2 Vol 1, 790 (1968)
14. G.T. Croft, F.J. Donahoe, W.F. Love, Rev. Sci. Instr., 26, 360 (1955)
15. W.D. Westwood, B. Bridger, E.L. Murray, J. Sci. Instr., Series 2 Vol 3, 962 (1970)
16. H.J. Condon and J.A. Marcus Phys. Rev., 134, A446 (1964)
17. G. Maxim, J. Sci. Instr. Series 2 Vol 2, 319 (1969)
18. A.P. King, G. Robinson, J.A. Cundall and M.J. Hight, J. Sci. Instr., 41, 766 (1964)
19. F.B. Humphrey and A.R. Johnston , Rev. Sci. Instr., 34, 348 (1963)
20. J. Vanderkooy, J. Sci. Instr., Series 2 Vol 2, 718 (1969)
21. C. Voigt and S. Foner J. Phys. E., (Sci. Instr.,) 5, 126 (1972)
22. F.B. Hagedorn and E.M. Gyorgy, J. Appl. Phys. 39, 995 (1968)

23. H. Shenker, "Magnetic anisotropy of cobalt ferrite"
Ph.D. thesis, University of Maryland.
Publication No. 12090, University Microfilms, Ann Arbor, Michigan, 1955

24. P.J. Flanders, Rev. Sci. Instr., 41, 697 (1970)

25. P.J. Flanders, J. App. Phys., 42, pp. 1635 (1971)

26. G.W. Rathenau and J.L. Snoek, Physica 8, 555 (1941)

27. H. Zijlstra, Rev. Sci. Instr., 32, 634 (1961)

28. J.C. Slonczewski, "Magnetism" Vol 1, p. 205, Ed. Rado and Suhl Academic Press, London and New York 1963

29. A. Braginski, Phys. Stat. Sol.,11, 603 (1965)

30. R.W. Teale, Proc. Phys. Soc., 92, 411 (1967)

31. A. Broese van Groenou, J.L. Page and R.F. Pearson
J. Phys. Chem. Solids, 28, 1017 (1967)

32. J.E. Knowles, J. Sci. Inst., Series 2 Vol 2, 917 (1969)

33. J.E. Knowles, J. Phys. E., (J. Sci. Instr.,) 7, 91 (1974)

34. V. Frank and O.V. Nielson,
J. Phys. E. (Sci. Instr.,) 4, 346 (1971)

35. P.K. Baltzer, R.C.A. Industrial Service Laboratory
Report No. RB-132 (1957)

36. W.P. Mason, Phys. Rev., 96, 302 (1954)

37. W.L. Bond, Rev. Sci. Instr., 22, 344 (1951)

38. P.H. Cross, Rev. Sci. Instr., 32, 1179 (1961)

39. Y.W. Lam, J. Sci. Instr., 42, 761 (1965)

40. J.R. Chamberlain, Proc. Phys. Soc., 78, 819 (1961)

41. G.F. Dionne, J. App. Phys., 40, 1839 (1969)

42. R.J. Cunningham, "Effect of non magnetic scandium octahedral substitution on the magnetization and initial permeability of YIG"
U.S. Naval Ordnance Laboratory, Technical Report 63-206 (1963)

43. R.A. Laudise, "The growth of single crystals" Prentice Hall Inc. Englewood Cliffs, 1970

44. B.R. Pamplin (Editor) "Crystal Growth" Pergamon Press, Oxford 1975.

Magnetostriction

E.W. LEE
University of Southampton

1. Introduction

The elementary magnetic moments in a crystalline solid interact not only with an external field and with each other but also with the crystal lattice. If, as is almost always the case, this interaction varies with lattice strain the magnetic moments will effectively exert a force on the lattice. This feature of the interaction is usually termed magnetoelastic coupling and is responsible for many features of the behaviour of magnetically ordered solids. It provides the means whereby a spin system in an excited state can transfer its excess energy to the lattice and so determines the spin damping which is responsible for the intrinsic line width of the ferromagnetic resonance and the closely related domain wall damping in ferro- or ferrimagnetic insulators. It determines the strength of the spin-phonon coupling which can be studied by the absorbtion of microwave phonons or by the study of the low temperature thermal conductivity of otherwise pure crystals doped with magnetic impurities. Another manifestation is the mixing of magnon and phonon modes which is experimentally observable by inelastic neutron scattering. These are all dynamic effects associated with the collective excitations of the coupled magnetic and crystal lattice.

The static effects of magnetoelastic coupling, which can be regarded as the long wavelength limits of all the dynamic effects, are usually denoted by the term magnetostriction. There are two principal effects. The first is a uniform lattice distortion (with respect to the paramagnetic phase) whose appearance coincides with the onset of magnetic order. The distortion can usually be decomposed into two parts. One of these depends only on the degree of magnetic order and it preserves the symmetry of the paramagnetic phase. The other depends on the degree of magnetic order and also on its direction within the lattice. This anisotropic distortion which does involve a loss of symmetry is usually smaller than the isotropic part, often by as much as two or three orders of magnitude. The second effect is a thermodynamic inverse of the first and appears as a stress-dependent contribution to the magnetic free energy. The isotropic part manifests itself as a dependence of the magnetic ordering temperature on external stress; the anisotropic part appears as a stress-dependent anisotropy whereby the axis of the magnetic ordering is determined by an external stress. This gives rise to technologically important effects such as the stress dependence of the technical magnetization curve; non-uniform internal stresses in real crystals or polycrystalline aggregates interact via the static magnetoelastic coupling to produce local variations in the magnetic free energy which place an upper limit on the initial susceptibility and a lower limit on the coercivity of the material.

A third effect which is really a combination of both the above effects is a dependence of the elastic constants on the direction and/or magnitude of the degree of magnetic order. This appears, for example, as an anomalous temperature dependence of the elastic constants in the neighbourhood of the ordering temperature, an affect which is exploited technically to produce alloys in which the temperature coefficient of an elastic constant (or of the speed of propagation of an elastic wave) is made very small over a finite temperature range.

All these static effects, and some of the dynamic ones as well, can be described in terms of a small number of magnetostriction or of magnetoelastic constants. In simple cases it is possible to calculate these from a knowledge of the electronic structure of the relevant ion. However close numerical agreement between the calculated and measured constants has hardly ever been achieved and the task of determining these constants is invariably entrusted to the experimenter. In this chapter we review the techniques available to him and discuss their sensitivity and the accuracy he is likely to achieve. In order to appreciate more fully what experiments can be performed it is necessary to outline the formal theory of magnetostriction in crystals.

2. Theory

The remarks of the previous section are perfectly general and the term "magnetic order" has been used throughout to emphasize that they apply to both ferromagnets and antiferromagnets. The formal theory of magnetoelastic effects is complicated by notational complexities. In an attempt to make the physical situation quite clear the theory is first developed for a one-dimensional system. In doing so we lose any distinction between magnetoelastic effects in antiferromagnetic and ferromagnetic systems. Extension to 3-dimensional ferromagnetic crystals is straightforward but it is less easy to generalise the results to antiferromagnetic ordering. This is therefore an appropriate stage to drop the pretence that both types of ordering can be treated within a general formalism; for the remainder of this chapter we shall be concerned only with ferromagnets. The theory is conventional and closely follows that of Carr[1]; for discussion and criticism of conventional theory the reader is referred to the books by Birss[2] and by Brown[2].

Magnetostriction occurs when there exists a contribution, of magnetic origin, to the free energy of a system which is linear in strain. This strain-dependent free energy is equivalent to a set of constant forces which deform the crystal until the forces are exactly opposed by the elastic forces. The "system" may be an isolated crystal, in which case the appropriate free energy is the Helmholtz free energy integrated, if necessary, over the volume of the crystal. Alternatively the "system" is a magnetic crystal either in a field of force or subject to a system of externally applied stresses in which case the appropriate free energy, integrated over the volume of the crystal, is the Gibbs free energy. Since we are not concerned with details such as the anisotropy of the magnetization it is permissible to ignore entropy changes. This effectively restricts the discussion to temperatures well below the magnetic ordering temperature. The Helmholtz free energy F is then written

$$F = F_m + F_e$$

where F_m and F_e are the magnetic and elastic parts of the free energy. It is assumed that F_m can be expanded as a Taylor series in the strain, e, and so

$$\begin{aligned} F_m &= F_o + F_1 + F_2 \\ &= F_o + V_oBe + \tfrac{1}{2} V_oB'e^2 \end{aligned}$$

Here V_o is the unstrained volume and B and B' are the first and second strain derivatives of the free energy. B is a stress, usually termed a magnetoelastic constant and B' is a magnetic contribution to the elastic stiffness constant. Similarly

$$F_e = \frac{1}{2} V_o C' e^2$$

where C' is the lattice elastic stiffness constant. It is usually assumed that $C' \gg B'$ and that both quadratic terms can be grouped together by setting $C = C' + B'$ where C is total elastic stiffness measured in an ordinary experiment, for example one in which the velocity of propagation of elastic waves is measured. Thus

$$F = F_o + V_o Be + \frac{1}{2} V_o Ce^2$$

The Gibbs free energy $G = F - V_o \sigma e$ where σ is the external stress. In this case it is convenient to decompose G into a magnetic and a lattice part as before but this time the natural variable is σ rather than e and the expansion takes the form

$$G_m = G_o + G_1 + G_2$$

$$= G_o + V_o A\sigma + \frac{1}{2} V_o A' \sigma^2$$

$$G_e = \frac{1}{2} V_o S' \sigma^2$$

Exactly as before A' is a magnetic contribution to the elastic modulus or compliance of the system, S' is the lattice elastic modulus and it is convenient to write $S = S' + A'$. The quantity A is a strain of magnetic origin. The external stress combines with the <u>total</u> lattice strain, e, to contribute a term $G_o = -V_o \sigma e$ to the Gibbs free energy. We therefore write

$$G = G_o + V_o (A-e)\sigma + \frac{1}{2} V_o S\sigma^2$$

Let us first consider the external stress to be zero. Then

$$F = F_o + V_o Be + \frac{1}{2} V_o Ce^2$$

$$G = G_o$$

The equilibrium, or spontaneous strains are obtained by setting $^{\partial F}/_{\partial e} = 0$. Thus

$$e_s = {}^{-B}/_C$$

$$= -BS$$

since in this model $S = {}^1/_C$. Hence the equilibrium value of F is

$$F^{eq} = F_o - V_oB^2S + \tfrac{1}{2} V_o CB^2S$$

$$= F_o - \tfrac{1}{2} V_oB^2S$$

$$= F_o - \tfrac{1}{2} V_oCe_s^2$$

This shows how the free energy of the system is lowered by allowing the spontaneous strain to occur. If an arbitrary strain e is imposed upon the system

$$F = F_o - V_oCe_se + \tfrac{1}{2} V_oCe^2$$

Next, suppose that an external stress is present. Then

$$G = G_o + V_o(A - e)\,\sigma + \tfrac{1}{2} V_oS\sigma^2$$

and the condition $^{\partial G}/_{\partial\sigma} = 0$ yields

$$e = A + S\sigma$$

This shows that the total strain can be considered to be the sum of an elastic strain due to the external stress and a magnetic strain independent of σ. If $\sigma = 0$ then $e = A = e_s$, the spontaneous strain previously shown to be equal to -BS. Thus A = -BS. The free energy in the presence of an arbitrary external stress σ is therefore

$$G = G_o + V_o (-BS-e)\sigma + \tfrac{1}{2} V_o S\sigma^2$$

or, in terms of the equilibrium strain,

$$G = G_o + V_o (e_s - e)\sigma + \tfrac{1}{2} V_o S\sigma^2$$

We note, also that F_o is the free energy at zero lattice strain. G_o if the free energy at constant lattice stress. The free energy at zero lattice strain can be obtained by putting the total strain equal to zero, i.e. $e = 0$ so that $A = -S\sigma_s = -BS$. Then

$$G = G_o - V_o S\sigma_s^2 + \tfrac{1}{2} V_o S\sigma_s^2$$

$$= G_o - \tfrac{1}{2} V_o S\sigma_s^2$$

so that

$$F_o = G_o - \tfrac{1}{2} S\sigma_s^2$$

$$= G_o - \tfrac{1}{2} SB^2$$

$$= G_o - \tfrac{1}{2} CA^2$$

$$= G_o + \tfrac{1}{2} AB$$

Thus a knowledge of the elastic constants and either A or B is sufficient to calculate all magnetoelastic effects to this order. In general it is easier to measure the equilibrium strains, e_s, rather than the stresses σ_s, required to annul them. However one can measure G as a function of an external stress and in certain cases, for example non-magnetic crystals containing small numbers of magnetic ions, this is the standard procedure.

To make contact with reality and, at the same time, to illustrate the complications imposed by real crystals we consider the example of a cubic ferromagnet. The stresses and strains become second rank tensors and the elastic constants and moduli become tensors of fourth rank.

The quantities F_m and G_m are made up of isotropic contributions to the free energy due to isotropic exchange and an anisotropic part due to single-ion (crystal field) and two-ion (e.g. dipolar) interactions and anisotropic exchange. Since the microscopic interactions leading to these anisotropic free energies are not properly understood it is customary to lump them all together. Then, since the proper angular dependence is also unknown, one expresses them as a function of the orientation of the total macroscopic magnetic moment relative to the crystal axes by means of a power series in either direction cosines or spherical harmonics. The latter have certain advantages, mainly of a mathematical nature; the former have greater practical utility. Denoting by α_1, α_2 and α_3 the direction cosines of the macroscopic moment vector with respect to the crystal axes we have

$$F_o = K_o' + K_1' (\alpha_1^2 \alpha_2^2 + \alpha_2^2 \alpha_3^2 + \alpha_3^2 \alpha_1^2) + K_2' \alpha_1^2 \alpha_2^2 \alpha_3^2 + \dots \quad (2.1)$$

The constants K_1' and K_2' are the anisotropy constants at zero strain. Since e is now a tensor of rank two the constant B divides into two types,

$$B_{11} = b_o + b_1 (\alpha_1^2 - \tfrac{1}{3}) + b_3 s + b_4 (\alpha_1^4 + \tfrac{2}{3} s - \tfrac{1}{3}) + \dots \quad (2.2)$$

$$B_{12} = b_2 \alpha_1 \alpha_2 + b_5 \alpha_1 \alpha_2 \alpha_3^2 + \dots$$

where $s = \alpha_1^2 \alpha_2^2 + \alpha_2^2 \alpha_3^2 + \alpha_3^2 \alpha_1^2$. Thus

$$\begin{aligned}
\frac{F_1}{V_o} = {} & b_o (e_{11} + e_{22} + e_{33}) + \\
& + b_1 \{ (\alpha_1^2 - \tfrac{1}{3}) e_{11} + (\alpha_2^2 - \tfrac{1}{3}) e_{22} + (\alpha_3^2 - \tfrac{1}{3}) e_{33} \} + \\
& + b_2 \{ \alpha_1\alpha_2 (e_{12} + e_{21}) + \alpha_1 \alpha_3 (e_{13} + e_{31}) + \alpha_2 \alpha_3 (e_{23} + e_{32}) \} + \\
& + b_3 s (e_{11} + e_{22} + e_{33}) + \qquad (2.3) \\
& + b_4 \{ (\alpha_1^4 + \tfrac{2}{3} s - \tfrac{1}{3}) e_{11} + (\alpha_2^4 + \tfrac{2}{3} s - \tfrac{1}{3}) e_{22} + (\alpha_3^4 + \tfrac{2}{3} s - \tfrac{1}{3}) e_{33} \} \\
& + b_5 \{ \alpha_1\alpha_2\alpha_3^2 (e_{12} + e_{21}) + \alpha_1\alpha_3\alpha_2^2 (e_{13} + e_{31}) + \alpha_2\alpha_3\alpha_1^2 (e_{23} + e_{32}) \} + .
\end{aligned}$$

This is known as the strain energy. It is the contribution made to the free energy by the strain e_{ij}. For a cubic crystal

$$\frac{F_2}{V_o} = \tfrac{1}{2} c_{11} (e_{11}^2 + e_{22}^2 + e_{33}^2) + c_{12} (e_{11}e_{22} + e_{11}e_{33} + e_{22}e_{33}) + \qquad (2.3)$$

$$+ \frac{c_{44}}{2} \quad (e_{12} + e_{21})^2 + (e_{13} + e_{31})^2 + (e_{23} + e_{32})^2$$

Similarly

$$G_o = K_o + K_1(\alpha_1^2\alpha_2^2 + \alpha_2^2\alpha_3^2 + \alpha_3^2\alpha_1^2) + K_2\alpha_1^2\alpha_2^2\alpha_3^2 + \ldots \qquad (2.4)$$

in which the constants K_1 and K_2 are the anisotropy constants at zero stress.

$$\frac{G_1}{V_o} = -h_o (\sigma_{11} + \sigma_{22} + \sigma_{33}) -h_1 \left\{(\alpha_1^2 - \tfrac{1}{3})\sigma_{11} + (\alpha_2^2 - \tfrac{1}{3})\sigma_{22} + (\alpha_3^2 - \tfrac{1}{3})\sigma_{33}\right\}$$

$$-2h_2\left\{\alpha_1 \alpha_2 \sigma_{12} + \alpha_1 \alpha_3 \sigma_{13} + \alpha_2 \alpha_3 \sigma_{23} - \ldots\right\}$$

This is known as the stress energy. It is the contribution to the free energy made by an external stress σ_{kl}. The spontaneous strains at zero stress are A_{ij}. These are related to B_{ij} as follows:

$$A_{ii} = - s_{11}B_{ii} - s_{12}(B_{ii} + B_{kk}), \; i = j = k$$

$$= h_o + h_1 (\alpha_i^2 - \tfrac{1}{3}) + h_3 s + h_4 (\alpha_i^4 + \tfrac{2}{3} s - \tfrac{1}{3}) + \ldots \qquad (2.5)$$

$$A_{ij} = - \tfrac{1}{2} s_{44}B_{ij} , \qquad i \neq j \neq k \qquad (2.6)$$

$$= h_2\alpha_i\alpha_j + h_5\alpha_i\alpha_j\alpha_k^2 + \ldots$$

Thus

$$h_o = -b_o(s_{11} + 2s_{12})$$

$$h_1 = - b_1(s_{11} - s_{12})$$

$$h_2 = -\tfrac{1}{2}b_2 s_{44}$$

$$h_3 = -b_3(s_{11} - s_{12})$$

$$h_4 = -b_4(s_{11} - s_{12})$$

$$h_5 = -\tfrac{1}{2} b_5 s_{44}$$

Also, for the cubic system

$$K_o = K_o' - \frac{3}{2} h_o^2 (c_{11} + 2c_{12}) + \ldots$$

$$K_1 = K_1' + h_1^2 (c_{11} - c_{12}) - 2h_2^2 c_{44} - 3h_o h_3 (c_{11} + 2c_{12}) + \ldots$$

$$K_2 = K_2' - 3h_1 h_4 (c_{11} - c_{12}) - 12h_2 h_5 c_{44} + \ldots$$

Thus we need to measure either the h's or the b's and to know either c_{ij} or s_{ij}. The number of independent h's or b's depends upon the degree of accuracy required. For many purposes an adequate approximation can be made by setting
$K_2 = K_2' = 0$, $h_3 = h_4 = h_5 = b_3 = b_4 = b_5 = 0$
In other cases further higher terms are needed. The form of equations such as 2.2 and 2.6 are dictated by the requirements of crystal symmetry and general expressions for all crystal classes have been given by Döring and Simon[3]. We shall not quote their general results but use their equations for special cases considered separately.

Although it is possible, in principle, to measure the b's directly a much easier experiment to perform is one which measures the spontaneous strains A_{ij}. If these are large enough it may be possible to measure certain of them using X-ray diffraction. In general, however, this is not so and the usual method is to measure the change of strain in a crystal as the direction of the macroscopic moment, hence α_i, is changed by the application of an external magnetic field. If the direction in which the strain is observed is specified by direction cosines β_1, β_2 β_3 the strain in this direction is

$$e(\alpha_i, \beta_j) = \sum_{i,j} A_{ij} \beta_i \beta_j \qquad (2.7)$$

Substituting for the A_{ij} from (2.6) one obtains, for cubic crystals,

$$e(\alpha_i,\beta_i) = h_o + h_1(\alpha_1^2\beta_1^2 + \alpha_2^2\beta_2^2 + \alpha_3^2\beta_3^2 - \tfrac{1}{3})$$

$$+ 2h_2(\alpha_1\alpha_2\beta_1\beta_2 + \alpha_2\alpha_3\beta_2\beta_3 + \alpha_3\alpha_1\beta_3\beta_1)$$

$$+ h_3 s \ldots\ldots\ldots\ldots\ldots\ldots (K_1 > O)$$

$$+ h_3(s-\tfrac{1}{3}) \ \ldots\ldots\ldots\ldots (K_1 < O)$$

$$+ h_4(\alpha_1^4\beta_1^2 + \alpha_2^4\beta_2^2 + \alpha_3^4\beta_3^2 + \tfrac{2}{3}s - \tfrac{1}{3})$$

$$+ 2h_5(\alpha_1\alpha_2\alpha_3^2\beta_1\beta_2 + \alpha_2\alpha_3\alpha_1^2\beta_2\beta_3 + \alpha_3\alpha_1\alpha_2^2\beta_3\beta_1)$$

It is conventional to refer to such strains as magnetostriction and it is the task of the experimenter to devise suitable techniques for measuring the strains $e(\alpha_i,\beta_j)$ from which the magnetostriction constants $h_o - h_5$ may be determined.

3. Experimental Method

3.1 General remarks

The measurement of strain (rather than strain differences) presupposes the existence of a well-defined and physically realisable state for which the strain can be taken as zero. In a purely mechanical system this presents little difficulty. The state of zero strain can be taken to correspond with that of zero stress. Of course the actual state of zero stress cannot be achieved exactly but provided the initial stress is small compared with the stresses that are eventually applied, as in a typical tensile testing experiment, the error involved is negligible. In magnetic systems this is by no means the case. The magnetic strains depend on the magnitude of the magnetic moment M and although the direction of M may be changed isothermally the magnitude of M can, in general, only be changed by allowing the temperature to change. This, inevitably, is accompanied by a change in lattice dimensions which is both magnetic and non-magnetic in origin and the latter is usually much greater than the former. The state of zero strain for magnetic systems is, in many cases arbitrary and is often a virtual state obtained by extrapolation from some non-magnetic state. This problem is most acute for the strains, such as h_o, which are independent of α_i. For this reason we defer consideration of the measurement of these strains and restrict our attention, for the present, to the determination of the α_i - dependent terms.

3.2 X-Rays

Consider, first, an isolated cubic magnetic crystal at zero stress. The equilibrium states of the crystal correspond to the minima of G_o. If, for the sake of simplicity, we set $K_2 = 0$ then the minima of G_o are:

(a) $K_1 > 0, \quad \alpha_i^2 = 1, \quad \alpha_j^2 = \alpha_k^2 = 0$

(b) $K_1 < 0, \quad \alpha_i^2 = \alpha_j^2 = \alpha_k^2 = \frac{1}{3}$

In the first case there are six solutions corresponding to domains, in each of which the magnetic moment is parallel to a cube edge. In the second case there are eight solutions corresponding to domains in which the magnetic moment is along a body diagonal. If we consider one domain of the first kind, for example that for which $\alpha_1 = +1, \alpha_2 = \alpha_3 = 0$ the spontaneous strains are

$$A_{11} = h_o + \frac{2}{3} h_1 + \frac{2}{3} h_4$$

$$A_{22} = A_{33} = h_o - \frac{1}{3} h_1 - \frac{1}{3} h_4$$

$$A_{12} = A_{23} = A_{31} = 0$$

This corresponds to tetragonal symmetry, the x-axis ($\alpha_1 = 1$) becoming the tetragonal c-axis with $^c/a = 1 + h_1 + h_4$, provided, as is always the case, that $(h_1 + h_4) << 1$. In a multi-domain crystal the c-axis will vary between the x,y and z axes and so the X-ray reflections, notably the (h00) reflections, will be split and the magnitude of the splitting yields $^c/a - 1$ directly. In a crystal for which $K_1 < 0$ the spontaneous strain in one domain, for example that for which $\alpha_1 = \alpha_2 = \alpha_3 = +\frac{1}{\sqrt{3}}$, is

$$A_{11} = A_{22} = A_{33} = \frac{1}{3} h_3$$

$$A_{12} = A_{23} = A_{31} = \frac{1}{3} h_2 + \frac{1}{9} h_5$$

From (2.7) this can be seen to represent a distortion to rhombohedral symmetry. The splitting of the X-ray reflections, notably those from (h,1,1) planes yields $\frac{4}{9} h_2 + \frac{4}{27} h_4$.

Although this permits what is, in many respects, the most direct measurements of the spontaneous strains the method suffers from several disadvantages. The sensitivity of X-ray methods is usually limited not by instrumental defects but by crystal imperfections. Therefore unless the spontaneous strains are very large or unless exceptionally perfect crystals are available the splitting of the Laue spots is not resolved and the method can do no more than indicate a distortion. The second disadvantage is of a more fundamental nature for it is evident that all that can be measured are linear combinations of certain constants. In principle, X-ray measurements could be performed with the sample in a magnetic field thus varying α_i and allowing all the constants to be determined but this does not appear to have been done.

The first magnetostriction measurement using X-rays was performed, accidentally, by Rooksby and Willis[4]. They measured the lattice parameter of Co-ferrite as a function of temperature and observed that, below the Néel temperature, the crystal symmetry was not cubic but tetragonal. At 80K the $^c/a$ ratio was determined as 0.99874. Rooksby and Willis took this to be evidence for an electronic transition such as occurs in magnetite. However, at this temperature $h_1 = -1.24 \times 10^{-3}$ [5] from which $^c/a = 0.99876$. Rooksby and Willis had measured the spontaneous magnetostriction. The technique was exploited by Darnell and Moore[6] who measured the distortion from hexagonal to orthorhombic symmetry as Dy passes from the helical to the ferromagnetic phase. These measurements, which cover the entire range of helical ordering are shown in Figure 1 to give some idea of the sensitivity and precision which can be obtained.

3.3 Displacement methods

There are not many materials whose magnetostriction is large enough to be readily measurable using X-rays. For most transition metals and alloys h_1 and h_2 are of the order 10^{-5} and thus approximately of the same magnitude as the strain due to a temperature change of 1K. The basic problem is not so much that of measuring a small strain or displacement but of ensuring that the temperature changes by no more than the tolerable limit during the time in which the measurement is made. Obviously this limit is set by the magnitudes of h_1 and h_2 and the precision required; in practice the limit is a few hundredths of a Kelvin.

The earliest measurements were made by clamping one end of a ferromagnetic rod or bar and measuring the displacement of the free end. The displacement is very small and of the enormous number of ingenious devices for magnifying the displacement all but one are of no more than historical interest. Contrary to what would, at first sight appear an obvious technique, methods based upon optical interference are rarely sensitive enough and, except in special cases where laser beams have been used, cannot compare with the simple combination of mechanical and optical lever first introduced by Nagaoka[7]. This is illustrated in Figure 2. The displacement is used to rotate a spindle on which is mounted a small mirror. A beam of light reflected off this mirror forms an optical lever whose movement can be measured by a number of standard methods. The methods is extremely simple and is both sensitive and accurate. Moreover it is absolute, in the sense that the required displacement can be found from easily measurable quantities.

Displacement methods suffer from two fundamental weaknesses. In the first place the displacement, and hence the sensitivity, is proportional to the size of the sample, a serious drawback when expensive or exotic crystals are being studied. Moreover the sensitivity cannot be usefully increased by the use of very long samples since this usually increases the difficulty of maintaining a uniform steady temperature. The second weakness stems from the fact that displacement methods are essentially axial or longitudinal in

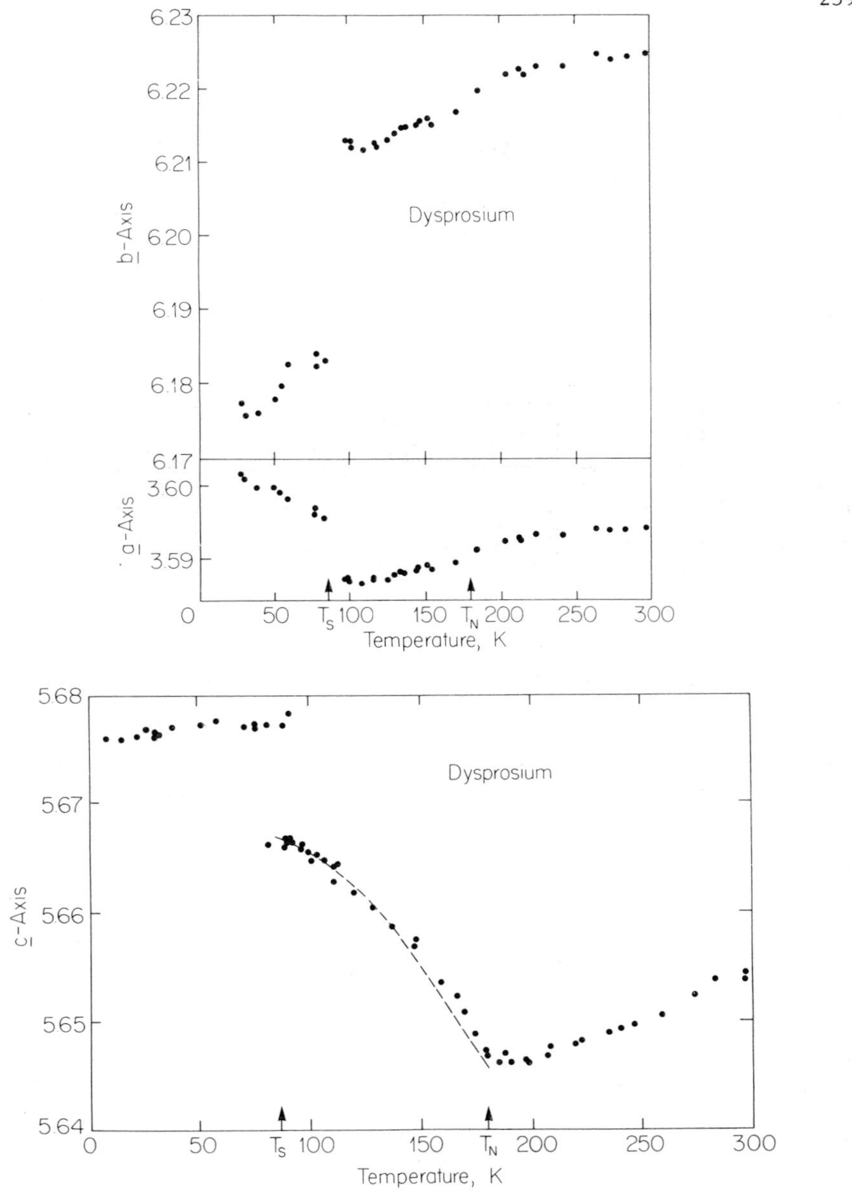

Figure 1. Lattice parameters of Dy as a function of temperature showing large magnetic contributions to the lattice expansion. The measurements were performed using Laue X-ray diffraction from small crystals. (Darnell and Moore[6])

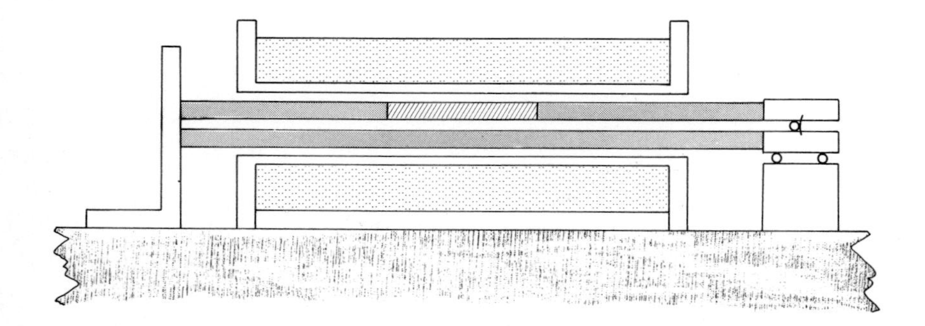

Figure 2 Simple apparatus for demonstrating magnetostriction in a rod-shaped sample. The sample is attached to end-pieces and placed in the centre of the solenoid. The lower horizontal rod affords some degree of temperature compensation. Sewing needles are suitable as spindles.

nature. They are not easily adapted to use with laboratory electromagnets although they are coming into their own again with the increasing use of high-power and superconducting magnets.

A further weakness is somewhat different in character though it has the same basic origin as the two mentioned above. A displacement method can really only yield a single measurement between two states. In order to attach a unique significance to the measurement both states must themselves be unique and well-defined. With this technique the two states which the experimenter is virtually obliged to use are the demagnetized multi-domain state for which the net magnetization of the crystal is zero and the saturated state in which the crystal consists, effectively, of a single domain. Provided the external field is applied along a symmetry axis of the crystal saturation is achieved in a finite field and provided that this is reached (and, for safety, exceeded by a generous margin) the saturated state is well-defined. However, there is no unique demagnetized state. The usual assumption is that equal volumes of the crystal have their domain magnetizations parallel to the six or eight easy directions. Thus if the strain is measured in a [100] direction for which $\beta_1 = 1$, $\beta_2 = \beta_3 = 0$ in a crystal for which $K_1 > 0$ the mean values of the angular functions are

$$\overline{\alpha_1^2} = \overline{\alpha_2^2} = \overline{\alpha_3^2} = \frac{1}{3}$$

$$\overline{\alpha_1\alpha_2} = \overline{\alpha_2\alpha_3} = \overline{\alpha_3\alpha_1} = 0$$

$$\overline{\alpha_1^4} = \overline{\alpha_2^4} = \overline{\alpha_3^4} = \frac{1}{9}$$

$$\overline{\alpha_i^2 \ \alpha_k\alpha_j} = 0$$

$$\overline{\alpha_1^2\alpha_2^2} + \overline{\alpha_2^2\alpha_3^2} + \overline{\alpha_3^2\alpha_1^2} = 0$$

The strain in this state

$$e_o = h_o - \frac{2}{9} h_4 \tag{3.1}$$

If the field is applied in the [100] direction, at saturation

$$\alpha_1 = 1, \alpha_2 = \alpha_3 = 0 \text{ and}$$

$$e_s = h_o + \frac{2}{3} h_1 + \frac{2}{3} h_4 \qquad (3.2)$$

Thus

$$e_s - e_o = \frac{2}{3} h_1 + \frac{4}{9} h_4 \qquad (3.3)$$

By performing several measurements varying, independently the direction of the field and that of measurement a series of values of $e_s - e_o$ can be obtained and from these the constants $h_1 - h_5$ can in principle be determined. Unfortunately the assumption of equal volumes upon which the average values in equation 3.2 are based has no backing, either theoretical or experimental. No proof has yet been offered that this is indeed the thermodynamically stable state and domain studies using direct observational techniques indicate that such a distribution is the exception rather than the rule. There is, in fact, an infinite number of domain distributions each of which leads to $\underline{M} = 0$ but to a different value of the mean strain in a crystal when the angular functions in equation 2.8 are averaged in accordance with each distribution. Except in special cases, which rarely appeal to the experimentalist because this usually means in crystals which are either very long or very thin, the domain distribution is not merely unknown but unknowable and so the strain in the demagnetized state is always uncertain. There may, of course, be circumstances in which the strain between the demagnetized and saturated states is itself of value but it should be clear that the constants $h_1 - h_5$ cannot be determined unambiguously from measurements made in this way.

3.4 The use of electrical resistance strain gauges

The resistance, R, of a conducting body depend upon its dimensions. For a wire of length l and of uniform cross sectional area A, $R = {}^{\rho l}/A$. When such a wire is stretched longitudinally the change in length, Δl, and the associated change in A combine to produce a change in resistance, ΔR, such that

$$\frac{\Delta R}{R} = (1 + 2\sigma) \frac{\Delta l}{l} \quad (3.4)$$

where σ is Poisson's ratio for the material and the resistivity is assumed to remain unchanged. The fractional resistance change is thus directly proportional to the strain in the wire. This is the principle of the electrical resistance strain gauge.

Strain gauges are commercially available in a wide variety of different forms and are extensively used by engineers to study static and dynamic strains in load-bearing and vibrating structures. Their application to the measurement of magnetostriction was pioneered by Goldman[8] and since that time they have almost completely supplanted all other methods. The literature on strain gauges is extensive and in this section only those features which have a direct bearing on magnetostriction measurement will be discussed.

The form of strain gauge widely used for the measurement of magnetostriction consists of a grid of metallic alloy embedded in or firmly attached to a matrix or base. In use, the base is fixed to the sample under investigation by an appropriate adhesive. The alloys which are used are usually based on conventional resistance alloys chosen for their high resistivity and low temperature coefficient. The physical form may be that of a fine wire looped to form a grid or a thin foil from which a grid is formed by photo-etching. Schematic diagrams of the form of these gauges, conventionally known as wire and foil gauges, respectively, are shown in Figure 3. The matrix of the wire gauges is usually hardened glue; the base may be paper or an epoxy-type film. In foil gauges the foil is usually deposited directly on to an epoxy backing. The physical characteristics of both types of gauge are remarkably constant and reproducible. Foil gauges are claimed to be better in this respect than the wire gauges. Smaller gauges can be made of the foil type than is possible with wire gauges and the latter are rapidly becoming obsolete.

In general, the resistivity of a metal is not independent of strain. Moreover, since resistivity and strain are both second rank tensors, they are connected by an elasto-resistivity tensor which is

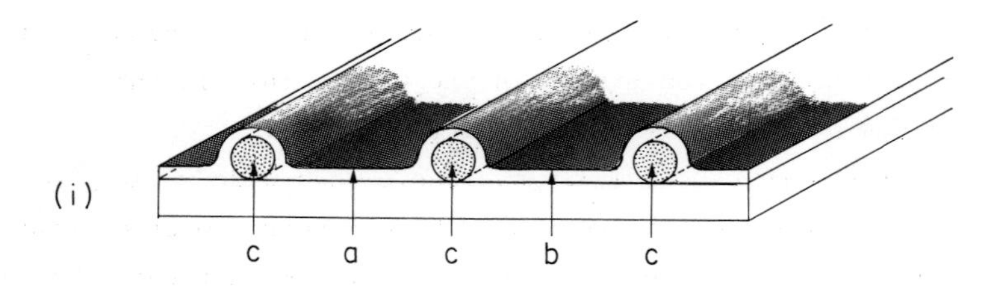

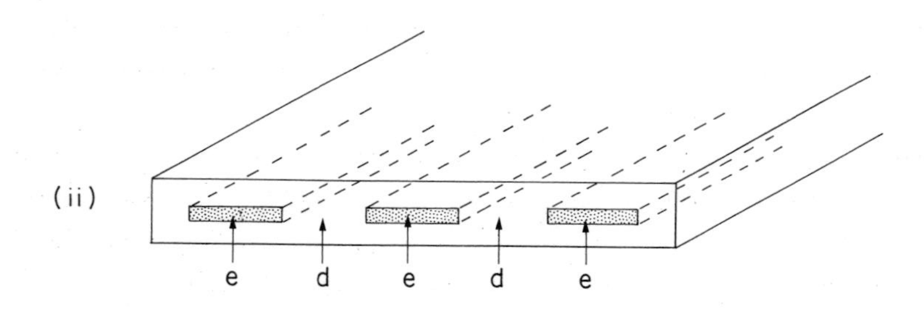

Figure 3 Schematic cross-section illustrating the construction of (i) wire and (ii) foil strain gauges.
a - paper backing, b - adhesive cement, c - strain gauge wire element, d - epoxy-type backing, e - strain gauge foil element.

of fourth rank. For such a tensor the number of independent components is never less than two, even for an isotropic solid. Therefore the usual assumption, that the resistivity of a metal is constant or is, at best, a function of its volume only, is without foundation. Since the elasto-resistance constants of resistance alloys are not known it is usual to write equation 3.4 in the form

$$\frac{\Delta R}{R} = G.\frac{\Delta l}{l} \tag{3.5}$$

in which the gauge factor, G, contains changes in R due to both geometrical and resistivity changes, and is regarded as a quantity to be determined by experiment. This is usually done by the manufacturer by fixing a gauge to a beam which is bent elastically thus imposing upon the gauge a calculable strain. Since the recommended adhesives are nearly always thermo-setting resins removal of the gauge usually involves its destruction. The gauge factor of a gauge is necessarily inferred from the consistency of the results obtained with other gauges of the same type. This is held to be sufficiently good to enable the manufacturer to quote the gauge factor with a precision of about 1%. Reproducibility under a set of fixed experimental conditions is one thing; the extent to which the quoted gauge factor may be relied upon to have the same value in a different set of conditions is what really determines the accuracy of a strain gauge and this does not seem to have been studied systematically. Apart from the nature of the adhesive (and it is clear that gauges should be fixed strictly in accordance with the manufacturers' instructions) the major source of uncertainty seems to arise from the change in resistance due to transverse strains. The behaviours of wire and foil gauges in this respect are likely to be different. When the member to which a wire gauge is fixed stretches in the direction of the gauge the strain is communicated to the wire via the backing and the matrix. The elongation of the wire is accompanied by a lateral contraction which is determined by the magnitude of the elongation and the elastic properties of the wire itself. From the nature of its construction, it seems unlikely that the accompanying transverse contraction of the underlying member is communicated to the gauge wire except at the ends of the grid where it contributes to the so-called transverse sensitivity - the chief objection to this type of gauge. In foil gauges the situation is rather different. Transverse sensitivity of the above-mentioned origin is reduced to negligible proportions by making the transverse sections much wider than the longitudinal ones, thus making little contribution to the total resistance. On the other hand the transverse strain in the longitudinal sections of the foil is now composed of a contribution from the foil metal and another imposed upon it by the transverse contraction of the underlying member to which it is fixed. For such gauges the measured gauge factor would appear to depend on the value of Poisson's ratio for the material to which it is fixed. Since most metals used for constructional purpose have similar Poisson's ratios of about 0.3 differences in the effective gauge factor are likely to go un-noticed. The effective Poisson's ratio for magnetostrictive strains associated with the change in direction of $\underline{M}$ is usually close to 0.5. Thus it is not immediately obvious that the manufacturers value for the gauge factor of a foil gauge may be assumed in this case. All one can do is point to the general consistency of measurements made with different gauges and to the absence of any report that measurement made on the same sample using different gauges produced systematically different results and to conclude that, through absence of any evidence to the contrary, it is safe to assume the manufacturer's quoted gauge factor.

The β - independent terms in equation 2.8 represent volume changes for which the effective Poisson's ratio is -1. The measurement of h_o presents special problems which are discussed more fully in Section 4. At this stage we merely remark that in this extreme case there seems to be little ground for the belief that the effective value of the gauge factor is identical with that obtained from measurements on a bent beam. The use of strain gauges for the measurement of magnetostriction in which substantial volume strains take place seems very much open to question.

The measurement of the resistance change is quite straightforward and is most conveniently accomplished using bridge methods. Numerous commercial bridges are available. Since magnetostrictive strains are often no larger than 10^{-5} the usual procedure is to measure the out-of-balance signal rather than to maintain the bridge at balance. A.C. bridges have the advantage of freedom from spurious voltages due to thermal e.m.f's. On the whole they are less sensitive than d.c. methods and have a poorer signal to noise ratio. D.C. bridges are simpler to set up; a sensitive galvanometer usually forms an adequate detector. Alternatively a commercial d.c. amplifier may be used to drive a more robust instrument or to operate a pen recorder. It is not difficult to achieve a sensitivity at which a strain of 10^{-7} can be measured with a precision of about 10%. This figure is usually set in practice both by the necessity to maintain constancy of temperature to about 0.01K and by the unwanted voltages induced by the changing external field.

Electrical resistance strain gauges can be used over a very wide range of temperature. They can all be operated at low temperatures and even the paper-backed wire gauges can be used successfully at liquid helium temperatures. The upper temperature limit is usually set by the nature of the backing and the bonding adhesive. Several foil gauges are claimed to be useful in the range 0 - 650K and special gauges are available for higher temperatures. However it must be remembered that ceramic cements are usually necessary for high temperature work and these may add appreciably to the stiffness of the material under investigation.

The gauge factor is temperature dependent and the form of the variation must be determined experimentally before undertaking measurements at different temperatures. For this purpose minature bending beam systems are used, mostly based on the design by McClintock[9].

A typical result is shown in Figure 4. Greenough and Lee[9] pointed out that the observed trend of G towards larger values at low temperatures is at variance with the likely temperature variation of Poisson's ratio. It might be connected with the variation of the elasto-resistivity coefficients which are likely to increase with decreasing temperature. However it is difficult to avoid the feeling that some, at least, of the temperature variation is connected with the properties of the backing and the adhesive.

The electrical resistance of pure metals and alloys can be changed by the application of a magnetic field. This

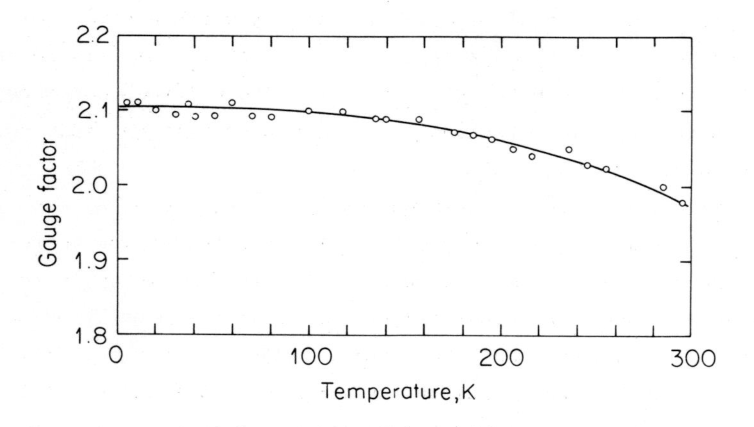

Figure 4 Gauge factor of a typical foil gauge as a function of temperature-data from F. Pourarian [9].

magnetoresistance effect is largest in pure metals and at low temperatures where the electronic mean free path is large compared with the cyclotron radius. In random substitutional alloys the disorder in the occupation of lattice sites reduces the mean free path to such a degree that the magnetoresistance is thought to be negligible. The resistance of alloys commonly used in strain gauges does not change by more than one or two per cent between room temperature and 4.2K. Nevertheless the magnetoresistance increases appreciably and is the major obstacle to the use of strain gauges for measurement of magnetostriction at low temperatures. Some results are shown in Figure 5. At very high fields the magnetoresistance sometimes tends to saturate and although this type of behaviour is not precluded in pure metals, it is difficult to avoid the suspicion that the metals themselves, nickel-based and containing other transition metals, are tending towards ferromagnetism at very low temperatures.

The exact composition of the metals used in strain gauges is not merely unknown but is often a closely guarded commercial secret. Slight variations in composition or the addition of further elements to improve other properties may have a profound effect on the degree of what may be termed "incipient magnetic order". It is therefore not surprising that the magnetoresistance behaviour of strain gauges is erratic, varying not merely from one type of gauge to another but often from one gauge to another within the same batch of nominally identical gauges. Fortunately the longitudinal and transverse magnetoresistances are often nearly equal so that in a measurement in which a magnetic field of constant magnitude is rotated about the sample the magnetoresistance tends to cancel. For other types of measurement no such cancellation may be assumed; the use of a dummy gauge situated in the same field as the sample but mounted on a non-magnetic base fails since there is no way of being certain that both gauges have the same magnetoresistance.

Finally, it should be observed that the application of a strain gauge to a sample will add to its stiffness, whilst the deforming force remains unchanged. The observed magnetostriction is likely to be lower than that of a free sample. Clearly, the reduction depends on the elastic properties of the gauge and its thickness relative to those of the sample under study. For transition metals and alloys the effect is small. The rare earth metals are elastically soft and in this case a correction may be necessary unless the samples are more than a few mm thick. The effect has been investigated for polycrystalline terbium with results[10] which are shown in Figure 6.

The above remarks are not meant to sound a cautionary warning but to draw attention to the vigilance which must be observed if accurate and reliable measurements of magnetostriction are to be obtained. Care is worth taking because the strain gauge method possesses two advantages which set it apart from all others. In the first place the strain is measured directly and so the sensitivity of the method is independent of sample size. Gauges are now available whose overall size is no more than 1.65 x 1.65 mm. Thus measurements are possible on exotic materials which are often only available as small crystals. Moreover the entire assembly can be placed inside a miniature cryostat. However the greatest advantage comes from the ability to measure strain differences when a magnetic field is rotated from one direction to another. The necessity to make use of a demagnetized state of unknown character disappears and all the uncertainties mentioned in section 3.3 vanish.

In a typical experimental arrangement the sample is a disc, usually between 6 and 10mm in diameter and about 2mm in thickness. A strain gauge is fixed to one side of the disc, the other side being mounted in such a way as to impose the least possible mechanical constraint to the strain in the disc; a suitable base is of cork containing a number of fine perpendicular saw-cuts so as to give it a "waffle-iron" appearance. A gauge from the same batch is mounted on a non-magnetic metal, preferably one whose thermal expansion is similar to that of the material under investigation. This dummy gauge should be mounted parallel to the active gauge and as close to it as possible. These two gauges form the ratio arms of Wheatstone's bridge. The sample plus dummy are usually then mounted in a suitable sample holder in a standard cryostat with their planes horizontal. The cryostat is then placed between the pole pieces of a laboratory electromagnet mounted on a rotating base. The measurement consists in measuring the change in strain as the direction of the magnetic field is altered. If the field is sufficiently large to convert the

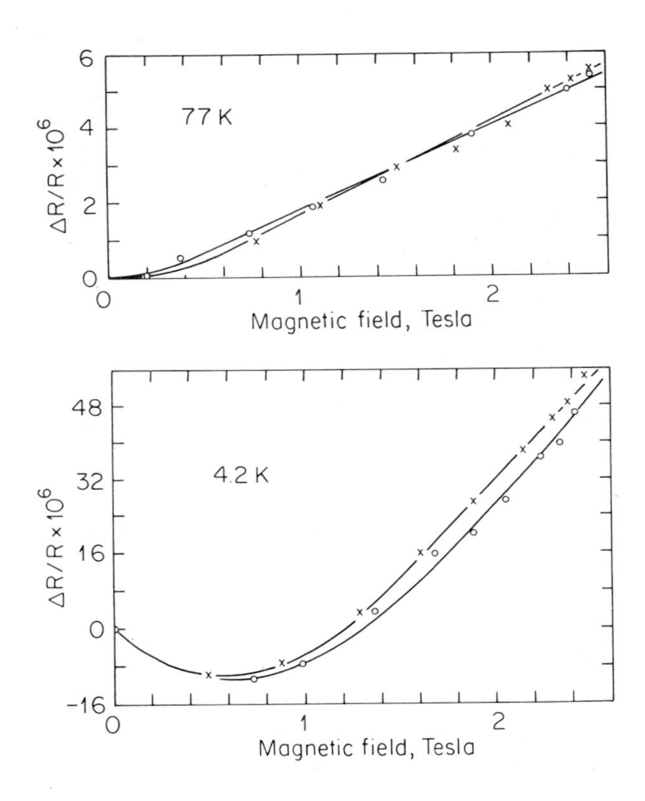

Figure 5 Magnetoresistance of a typical foil gauge at 77K and at 4.2K - data from F. Pourarian [9].

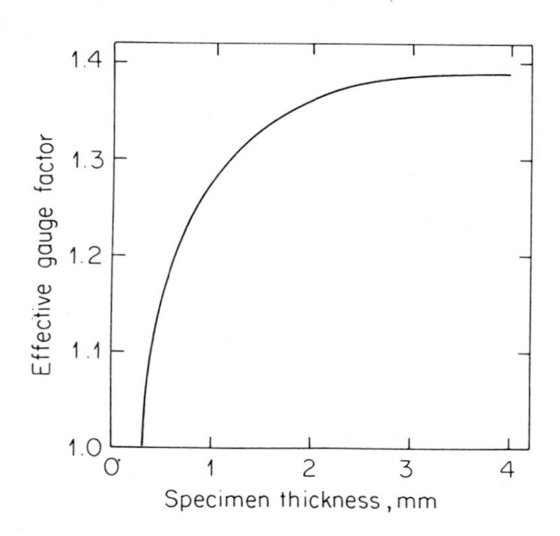

Figure 6 Effective gauge factor of a strain gauge with a bakelite backing as a function of the thickness of a disc of polycrystalline terbium. All the rare earth metals are elastically very soft and the restraining effect of a bonded gauge is relatively large.(du Plessis[10])

sample into a single domain with $\underline{M}$ parallel to the field direction α_i is determined. The direction of the gauge elements relative to the crystal axes defines β_i and so, in principle, one can always measure enough strains to determine as many of the constants h_n as seem to be required.

For cubic crystals the most useful plane is the (110) since from measurements in this plane the constants $h_1 - h_5$ can be determined. Let θ denote the angle between $\underline{M}$ and the [001] direction and suppose the gauge to be fixed parallel to [001] so that $\beta_1 = \beta_2 = 0$, $\beta_3 = 1$.

The strain is

$$e_\theta = h_1 \ (1 + 3 \cos 2\theta)/6 + h_3(7 - 4 \cos 2\theta - 3 \cos 4\theta)/32$$
$$+ h_4(9 + 20 \cos 2\theta + 3 \cos 4\theta)/48$$

Thus $$e_o = \frac{2}{3} h_1 + \frac{2}{3} h_4$$

and $$e_\theta - e_o = h_1(-1 + \cos 2\theta)/2 + h_2.0$$
$$+ h_3(7 - 4 \cos 2\theta - 3 \cos 4\theta)/32 \qquad (3.5)$$
$$+ h_4(-23 + 20 \cos 2\theta + 3 \cos 4\theta)/48 + h_5 \ . \ 0$$

Next, suppose the gauge is fixed parallel to [111] . Then

$$e_\theta - e_o = h_1.0 + h_2(1 - \cos 2\theta + 2\sqrt{2} \sin 2\theta)/6$$
$$+ h_3(7 - 4 \cos 2\theta - 3 \cos 4\theta)/32 + h_4.0 \qquad (3.6)$$
$$+ h_5(1 + 2\sqrt{2} \sin 2\theta - \sqrt{2} \sin 4\theta - \cos 4\theta)/24$$

To calculate θ we express the magnetic energy of the crystal as the sum of the anisotropy and field energies. Thus

$$E = K_1(7 - 4 \cos 2\theta - 3\cos 4\theta)/32 + K_2(2-\cos 2\theta - 2\cos 4\theta + \cos 6\theta)/128$$
$$- BM \cos (\theta - \phi)$$

where ϕ is the angle that the external field H makes with [001] and θ is given by $\partial E/\partial\theta = 0$. Thus

$$K_1\ (2\sin 2\theta + 3\sin 4\theta)/8 + K_2\ (\sin 2\theta + 4\sin 4\theta - 3\sin 6\theta)/64$$

$$+\ BM\sin(\theta - \phi) = 0$$

In the experiment one can measure ϕ but $(\theta - \phi)$ does not, in general, vanish for finite fields. However, if K_1, K_2 and M are known θ can be calculated from a knowledge of B and ϕ. Fortunately the correction, $\theta-\phi$, is quite insensitive to the value of K_2[11].

From this point the exact procedure is determined by the accuracy required. As an example we outline the routine devised by Bozorth and Hamming[12]. This involves calculating the strain differences $e - e_o$ at 5° intervals of θ from measurements at 5° intervals of ϕ as indicated above. This yields 18 strain differences for each gauge position and thus 36 equations relating strain differences to the constants $h_1 - h_5$. However the equations are neither independent nor mutually consistent since each strain measurement is liable to be in error by a finite amount. Bozorth and Hamming mix the two sets of measurements, minimize the sum of the squares of the errors with respect to different h's to obtain consistent equations

$$h_j = \sum_{ji} \delta_{ji}\ e_i$$

where j runs from 1 to 5 and i runs from 1 to 36 so that, for example, $e_3 = 3_{15^o} - e_o$ with the gauge parallel to [001] . The coefficients δ_{ji} are given in the paper cited above.

The method assumes that the two gauges used have identical characteristics. If they do not, then $e_1 - e_{18}$ may not be mixed with $e_{19} - e_{36}$. However since two gauge positions must be used there is little point in treating the two sets of strains independently since if the two sets of strain measurements are mutually inconsistent there is no way of telling which set is correct.

This is by no means the only possible procedure and there may be other ways of measuring the strains which could be advantageous, for example, if one of the constants is abnormally large or small.

Gersdorf[13] has described a method in which two strain gauges are fixed to the sample and are perpendicular to each other. When devising gauge configuration to suit special cases one must take care to ensure that small errors of misalignment do not make disproportionate errors in the constants. Misalignment is of two kinds;

(1) misalignment of crystal axes with respect to the plane of the crystal

(2) misalignment of gauge with respect to crystal axes.

These can be regarded as independent provided the degree of misalignment is small. Each crystal plane and each gauge configuration requires separate treatment and a general and detailed general analysis does not appear to have been made. Some special cases have been considered by Carr and Smoluchowski[14] and by Greenough[15].

The elaborate procedure outlined above is hardly worth pursuing unless the strain measurements themselves are almost free of error. A series of eighteen strain measurements takes several minutes to accomplish and , during this time, the temperature must remain constant to above 0.01K. In practice this is difficult to achieve except at room temperatures and at the normal boiling points of nitrogen and helium. At intermediate temperatures one often has to accept a procedure which yields less detailed information. For example consider equation 3.5. Denote the strain when M is parallel to the gauge by $\lambda_{\parallel}$ and when it is perpendicular to it by $\lambda_{\perp}$. Then

$$\lambda_{\parallel} - \lambda_{\perp} = e_{o} - e_{90} = h_{1} - \frac{1}{4} h_{3} + \frac{5}{6} h_{4}$$

When the gauge is along [111] $\lambda_{\parallel}$ corresponds to $\theta = \cos^{-1} 1/\sqrt{3}$ and

$$\lambda_{\perp} \text{ to } \theta = \cos^{-1} \left(\frac{1}{\sqrt{3}} + \frac{\pi}{2}\right). \quad \text{Then}$$

$$\lambda_{\parallel} - \lambda_{\perp} = h_{2} + \frac{1}{12} h_{3} + \frac{2}{9} h_{5}$$

Thus two pairs of strain differences measured in a (110) plane serve to determine two linear combinations of constants. If the magnetostriction is adequately described by the first two constants so that $h_3 = h_4 = h_5 = 0$ then the two strain differences can be regarded as yielding the two second order constants h_1 and h_2. The strain difference $\lambda_{\parallel} - \lambda_{\perp}$ can be measured in a few seconds and so h_1 and h_2 can be determined accurately at all temperatures. Other crystal planes and other gauge orientations may be used.[13,14] However the arrangement just described is particularly advantageous since, for all four strain measurements,

(1) the external field is parallel to an axis of high symmetry and so the magnetization is parallel to the field

(2) the strain is an extremum and is relatively insensitive to slight misorientation of the external field.

The next most important crystal symmetry is hexagonal since cobalt and the heavy rare earth metals come into this category. Several equivalent expressions for the strain in hexagonal crystals have been used. A typical one is

$$e = \lambda_o + \lambda'_o\beta_3^2 + (\lambda_1 + \lambda_2\beta_3^2)\,\alpha_3^2 + \lambda_3\,(\alpha_1\beta_1 + \alpha_2\beta_2)^2 + \\ + \lambda_4(\alpha_1\beta_1 + \alpha_2\beta_2)\,\alpha_3\beta_3 + (\lambda_5 + \lambda_6\beta_3^2)\,\alpha_3^4 + \lambda_7\,(\alpha_1\beta_1 + \alpha_2\beta_2)\,\alpha_3^2 + \\ + \lambda_8(\alpha_1\beta_1 + \alpha_2\beta_2)\,\alpha_3^3\beta_3 + \lambda_9\,(2\alpha_1\alpha_1\beta_1 + \alpha_1^2\beta_2 - \alpha_2^2\beta_2)^2 + \qquad (3.7)$$

in which the subscript 3 refers to the hexagonal c - axis. Cylindrical symmetry persists up to and including the term λ_8. The second order constants $\lambda_1 - \lambda_4$ are, in principle of equal importance and most work on simple hexagonal ferromagnets has been limited to the determination of these four constants.

Although it would appear possible in principle to determine all four constants from measurements in plane of low symmetry the usual procedure is to use at least two crystals each with a high-symmetry crystal plane. The magnetostriction of hexagonal crystals tends to be greater than for cubic crystals. The detailed analyses which have been made for cubic crystals have evidently been deemed neither necessary nor desirable for hexagonal crystals and for further details reference should be made to the work on Co[16], Gd[17], MnBi[18] and $BaFe_{18}O_{27}$[19].

The magnetostriction of the heavy rare earth metals is very large but these elements present special problems because the anisotropy is so large that ordinary laboratory fields are not strong enough to change the direction of M. Once again, reference should be made to the original papers [20].

3.5 Measurements using polycrystalline samples

The preparation and orientation of single crystals is, at best, a tedious task. Polymorphic transitions and, in the case of alloys, vagaries of the phase diagram often conspire to prevent the growth of usable crystals altogether. Such cases oblige the experimenter to work with polycrystalline samples and it is worth while enquiring what information measurements of such samples can be made to yield.

In the first place all the warnings uttered in Section 3.3 concerning non-random domain orientation and the danger of using the demagnetized state apply with even greater force to measurements on polycrystals. Almost all the early measurements, often made on drawn or cast rods, are unreliable for this reason. To this must be added the knowledge of the grain orientation distribution. If this is random, with no preferred orientation of cubic crystal axes then the strain can be expressed as

$$\lambda = A + B\cos^2\theta \qquad (3.8)$$

where

$$A = h_o - \frac{2}{15}h_1 - \frac{1}{5}h_2 + \frac{1}{5}h_3 - \frac{12}{105}h_4 - \frac{1}{35}h_5$$

and

$$B = \frac{2}{5}h_1 + \frac{3}{5}h_2 + \frac{12}{35}h_4 + \frac{3}{35}h_5$$

and θ is the angle between the direction of M and that along which the strain is measured. Denoting λ_o and λ_{90} by $\lambda_{\parallel}$ and $\lambda_{\perp}$ as before it is clear that

$$\lambda_{\parallel} - \lambda_{\perp} = B$$

The constant A contains h_o which is not measurable by any of the methods so far described. If we disregard it and if (and only if) $h_3 = 0$, then $A = -\frac{1}{3}B$. Equation 3.8 is then often written

$$\lambda = \frac{3}{2}\lambda_s \left(\cos^2\theta - \frac{1}{3}\right) \qquad (3.9)$$

At this point it is customary to describe the magnetostriction in terms of the two first order constants h_1 and h_2 only to regard the constant λ_s as being defined by

$$\lambda_s = \frac{2}{5} h_1 + \frac{3}{5} h_2$$

The best way of measuring λ_s is to apply a saturating field and to measure the strain when the magnetization is rotated from a position perpendicular to the direction of measurement to a position parallel to it. The strain difference from equation 3.9 is

$$\lambda_o - \lambda_{90} = \frac{3}{2}\lambda_s$$

independent of preferred domain orientation but not of preferred crystallite orientation. The quantity λ_s is a useful approximation when h_1 and h_2 are of the same sign. Otherwise one really needs the two separate constant and the pretence that λ_s is a meaningful quantity can lead to conclusions which are absurd, as, for example, when $|h_1|$ and $|h_2|$ are both large and $|h_1| = -\frac{3}{2}|h_2|$.

Two methods have been used to determine h_1 and h_2 from measurements on polycrystals. When h_1 and h_2 are of opposite sign it is often possible to decompose by visual inspection the strain versus external field curve, when a substance is magnetized from the demagnetized state to saturation, into that part which is associated with movement of the domain boundaries and that which is associated with the rotation of the magnetic moment towards the field direction[21]. Because it utilizes the demagnetized state the results suffer from all the uncertainties of that state.

The second method employs the result that, close to saturation the magnetostriction can be expressed as a power series in the applied field (i.e. after application of an appropriate correction for the demagnetizing field which, in this field range can be considered constant). The strain is[22]

$$e = \left(\frac{4}{15} + \frac{16}{105h} - \frac{64}{3003h^2}\right)G\ h_1 + \left(\frac{2}{5} - \frac{64}{105h} + \left(\frac{64}{3003} - \frac{16}{105}\right)\frac{1}{h^2}\right)G\ h_2 \ldots \tag{3.10}$$

where $h = {}^{BM}/K_1$ and G is a factor which takes into account the effect of magnetic interactions between the crystallities.

Since the coefficient of the leading term involves $(h_1 - h_2)$ measurements of this quantity can be combined with measurements of λ_s to yield h_1 and h_2 separately. This was first done by D'Yakov[23] who, however, claimed that the factor G has to be omitted from equation 3.10 in order to yield values of h_1 and h_2 for nickel which are in good agreement with those derived from measurement on single crystals. This is curious since the factor G is known to be important in the corresponding expression for magnetization. The matter is still not resolved.

The above remarks apply only to polycrystalline samples of cubic materials. For hexagonal materials the position is much more complicated. In the first place the first four constants in equation 3.7 are of equal importance. Secondly, although the magnetostriction of a polycrystalline material can be put in the form of equation 3.7 this does not in general reduce to equation 3.8 for a hexagonal material. Measurements on polycrystalline samples can do no more than yield linear combinations of the constants in equation 3.7.

4. The Measurement of h_o

The quantity of h_o and similar quantities represent strains of magnetic origin but which do not depend on the direction of the magnetization within the crystal. Such quantities can only be measured by making use of their dependence on the magnitude of the magnetization. This can be changed on a variety of ways but the only two which are of importance are by applying a large external field or by changing the temperature. The increase in the intrinsic magnetization of a ferromagnet due to an external field can arise through a number of independent mechanisms. The nature of these is usually disregarded and the increase in magnetization as a whole is termed the paraprocess. All the constants h_o - h_5

have a temperature dependence which is formally accounted for by requiring each constant to scale as some function of the magnitude of the spontaneous magnetization. Since this is, in fact, field dependent it follows that each of the constants $h_o - h_5$ is field dependent at high fields due to the paraprocess. This dependence is termed the forced magnetostriction. For most substances h_o is at least two orders of magnitude greater than h_1 or h_2 and so the forced magnetostriction is dominated by the field dependence of h_o. In a molecular field model the volume strain $\omega = 3h_o = k_o M_s^2$ where M_s is the spontaneous magnetization and k is a constant which is directly proportional to the volume derivative of the molecular field constant. Thus

$$d\omega/dH = 2M_s k_o \, dM_s/dH = \frac{6h_o \chi_{HF}}{M_s} \qquad (3.11)$$

where χ_{HF} is the high field or paraprocess susceptibility. If χ_{HF} can also be measured a determination of $d\omega/dH$ will yield h_o.

In the paramagnetic region above the ordering temperature and in paramagnetic systems generally the measured magnetostriction is proportional to H^2. Thus $\omega = 3h_o = k\,\chi^2 H^2$ where χ is the paramagnetic susceptibility. Since χ is considerably easier to measure than χ_{HF} this is, in many ways, an easier procedure to adopt. Unfortunately, even in fields of 10T the strain is usually quite small and special techniques have to be adopted to measure them. Clearly the extraction of h_o from the measured strains requires the assistance of a specific model and it is by no means clear that the two procedures outlined above, one using measurements in the paramagnetic phase, the other in the ferromagnetic phase, will yield the same value for k and hence, h_o.

The other way of changing the spontaneous magnetization is to allow the temperature to change and in this case h_o and similar quantities manifest themselves as a contribution to the thermal expansion. Thus by measuring the total thermal expansion of substance over an appropriate temperature interval and subtracting that part of it which is due to the ordinary lattice expansion the magnetic thermal expansion can be determined. In practice this is a good deal more difficult than it sounds especially when the ordering temperature is comparable with the Debye temperature. Measurements of the total thermal expansion over a wide temperature range are subject to systematic errors which are quite difficult to uncover. The lattice expansion cannot be calculated with certainty; and so one has all the usual magnification of uncertainties arising from the subtraction of two nearly equal quantities. These problems become much less acute when the ordering temperature is very much lower than the Debye temperature. In this case most of the total expansion is magnetic in origin; the lattice expansion can then be regarded as a small correction and in extreme cases is negligible.

Experimental techniques which have been used to measure thermal expansion include the use of X-rays, strain gauges and devices based on the direct measurement of displacement. The former is invaluable when the transition from a paramagnetic to a magnetically ordered state is accompanied by a change in crystal symmetry, for example when the ordered phase is antiferromagnetic. It is ordinarily not very sensitive however and certain precuations have to be taken to avoid systematic errors. Although strain gauges have been used to measure thermal expansion no systematic study of the limitations of the method has been made. The technique has the advantages of a sensitivity about two orders of magnitude greater than X-ray methods and moreover is independent of sample size. It has proved invaluable in the study of certain magnetic phase transitions, notably those which occur in chromium,[24,25] but its accuracy and reliability in quantitative determinations over an extended temperature range is at present unproven. Mechanical displacement techniques are simple and the principle of the optical lever can be exploited to give high sensitivity. The mechanical and optical magnification can be extended using split photo-electric or photo- emissive devices and systems based on these principles can be made which are capable of detecting displacements of 100 ppm.[26] They have the advantage of being usable over a wide temperature range with no significant change in sensitivity.

For measurements requiring extreme sensitivity over a limited temperature range the most widely used method employs a three-terminal capacitance bridge using ratio-transformers. This arrangement is free from problems due to unstable lead capacitance in the measuring circuit [27]. The exceptional sensitivity of the technique derives from the high sensitivity of commercially available bridges - typically 1 in 10^7 combined with the usual sensitivity of a parallel plate capacitor to a change in separation of its plates. If the plate separation is 10^{-3} cm and changes in the capacitance can be measured to 1 part in 10^7 this corresponds to a change in plate separation of 10^{-10} cm. Capacitance cells suitable for the measurement of thermal expansion have been described by White[28] and by Fawcett[29] for the measurement of magnetostriction.

An ultra high-sensitivity dilatometer operating at a microwave frequency of about 10^{10}Hz has been developed by Pudalov and Khaikin [30]. Under optimum conditions the limit of detection corresponds to strains of about 5×10^{-13} but the long term stability is poor and the useful limit of sensitivity is about 10^{-11}.

Several optical techniques making use of laser beams are known to have been developed but in most cases the sensitivity fall some way below that of the three terminal capacitor.

5. Miscellaneous Topics

5.1 Forced magnetostriction

The field dependence of the magnetostriction constants arises from the small but finite susceptibility above saturation. At first sight it would appear that forced magnetostriction constants could be determined by measuring the magnetostriction at high field strengths in appropriately closen crystallographic directions. However this type of measurement, in which the direction of measurement is held constant and the field strength is varied will only yield linear combinations of the required constants. Suppose we truncate equation 2.8 retaining only the first three terms so that

$$e(\alpha_i,\beta_i) = h_o + h_1 (\alpha_1^2\beta_1^2 + \alpha_2^2\beta_2^2 + \alpha_3^2\beta_3^2 - \frac{1}{3}$$

$$+ 2h_2 (\alpha_1\alpha_2\beta_1\beta_2 + \alpha_2\alpha_3\beta_2\beta_3 + \alpha_3\alpha_1\beta_3\beta_1) \qquad 5.1$$

Along [100] $\quad \frac{d}{dB} e(100,100) = h_o' + \frac{2}{3} h_1' \qquad 5.2a$

Along [110] $\frac{d}{dB} e(110,110) = h_o' + \frac{1}{6} h_1' + \frac{1}{2} h_2'$ 5.2b

Along [111] $\frac{d}{dB} e(111,111) = h_o' + \frac{2}{3} h_2'$ 5.2c

where the primed quantities are the required forced magnetostriction constants (e.g., $h_1' = \frac{dh'}{dB}$). Equations 5.2a - 5.2c are not linearly independent and so a measurement of the three quantities on the left will yield only linear combinations of h_o', h_1' and h_2'. To separate these at least one additional measurement must be made in which the field and the direction of measurement are not co-linear. For example

$$\frac{d}{dB} e(100,100) - \frac{d}{dB} e(100,010) = h_1'$$

$$\frac{d}{dB} e(100,100) + 2\frac{d}{dB} e(100,010) = 3h_o'$$

In transition metals measurements of $\frac{de}{dB}$ always include a contribution from h_o' which is often much larger than h_1' and h_2' and may also include an unwanted contribution from magnetocaloric heating unless precuations are taken to ensure isothermal conditions. The best procedure is, therefore, to perform the rotation type of measurement outlined in Section 3.4 at several different field strengths and to analyse the measurements taking into account contributions from the forced magnetostriction. At very low temperatures where the high field susceptibility is usually very small the forced magnetostriction is so small that the sensitivity of the three terminal capacitor technique becomes necessary.

5.2 Measurements under external stress

Instead of measuring the equilibrium strains the same information can be obtained from a measurement of the stress dependence of the magnetic fee energy. Static torque measurements under hydrostatic pressure can be performed [31]. In cubic crystals $\frac{dK_1}{dp}$ is directly related to the magnetostriction constant h_3 which is difficult to measure directly. Measurements relating to h_1 and h_2 require the application of axial stress and in this case the usual procedure is to determine the shift in the ferromagnetic resonance frequency under uniaxial compression. Measurements made in this way are said to be in good agreement with the measurement of equilibrium strains[32]. No additional information is obtained this way and the sensitivity and the precision of the method is usually inferior to that of direct measurement. Even for very small or very thin (e.g. epitaxial film) crystals the resonance is usually quite strong and in these cases the technique comes into its own.

5.3 Measurements in intense magnetic fields

Fields up to about 12T can be conveniently produced using superconducting magnets (though few are available with transverse access) and the measurement of magnetostriction in such fields can be performed using any of the techniques described in previous sections. The magnetoresistance of certain strain gauges at low temperatures may prove excessive when the strains to be measured are small; otherwise there are no additional problems.

Fields greater than about 20T can only be obtained for short times. The 1s pulse duration in the installation at the University of Amsterdam is unusual. In the usual pulsed field systems using capacitors for storing electrical energy, the field is typically a half sine wave of duration 10 ms. Measurements, using such fields, require special techniques. The pioneering work of Kapitza [33] was done using a purely mechanical sensor based on the mechanical force balance which had been successfully used as a magnetometer. This device was capable of measuring displacements of 10^{-7} cm. Using samples between one and two cm long Kapitza measured the linear magnetostriction of bismuth, antimony and graphite at room temperature and at the temperature of liquid nitrogen. Later methods have tended to make use of capacitive transducers[34] or electrostrictive transducers[35].

Although these methods are sensitive they suffer from all the disadvantages of displacement methods discussed in Section 3.3. It has been shown By Ricodeau et al[36] that electrical resistance strain gauges may be operated under pulsed field conditions using commercially available equipment. Magnetoresistance and electromagnetic pick-up is still a problem but the method can be used satisfactorily for the measurement of strains of the order 10^{-5}.

References

1. W.J. Carr, Handbuch der Physik, XVIII/2. (Berlin:Springer. 1966)
2. R.R. Birss, Electric and Magnetic Forces (London: Longmann 1967).
   W.F. Brown Jr., Magnetoeleastic Interactions (Berlin: Springer, 1966)
3. W. Döring and G. Simon, Ann. Physik, 5, 373 (1960)
4. H.P. Rooksby and B.T.M. Willis, Nature, 172, 1054 (1953)
5. E.W. Lee and J.A. Robey, Proc. Int. Conf. Magnetism, Nottingham 1964 (London: I.O.P.) 642
   R.D. Greenough and E.W. Lee, J. Phys. D, 3, 1595 (1970)
6. F.J. Darnell and E.P. Moore, J. Appl. Phys., 34, 1337 (1963).
   See also F.J. Darnell, Phys. Rev., 130, 1825 (1963),
   Phys. Rev., 132 128 (1963)
7. H. Nagaoka, Phil. Mag., 37 131 (1894)
   for a useful survey of displacement methods see L.F. Bates, Modern Magnetism, 4th Ed. (Cambridge: University Press 1961).
8. J.E. Goldman, Phys. Rev., 72, 529 (1947)
9. R.M. McClintock, Rev. Sci. Inst., 30, 715 (1959)
   R.D. Greenough and E.W. Lee, Cryogenics, 7, 10 (1967)
   F. Pourarian, Ph.D. thesis, University of Southampton (1973).
10. P. de V. du Plessis, Phil. Mag., 18, 145 (1968)
11. E.W. Lee and M.A. Asgar, Proc. Roy. Soc. Lond., A326, 73 (1971)
12. R.M. Bozorth and R.W. Hamming, Phys. Rev., 89, 865 (1953)
13. R. Gersdorf, Thesis, University of Amsterdam (1961)
14. W.J. Carr and R. Smoluchowski, Phys. Rev., 83, 1236 (1951)
15. R.D. Greenough,Thesis, University of Sheffield (1966)
16. R.M. Bozorth, Phys. Rev., 96, 311 (1963)
17. R.M. Bozorth and T. Wakiyama, J. Phys. Soc. Japan, 18, 97 (1963)
18. H.J. Williams, R.C. Sherwood and O.L. Boothby, J. Appl. Physics., 28, 445 (1957)
19. S.S. Fonton and A.V. Zalesskii, Soviet Physics, J.E.T.P., 20, 1138 (1965)
20. For a summary and an extensive list of references see the chapter by J.J. Rhyne in "The magnetic properties of rare-earths" ed. by R.J. Elliott, (Plenum Press, 1972).
21. M. Yamamoto and R. Miyasawa, Sci. Rep., RITU, A5, 113 (1953)
22. E.W. Lee, Proc. Phys. Soc. A, 67, 381 (1954)

23. G.P. D'Yakov, Fiz-metal metallovd., 6, 168 (1958)

24. E.W. Lee and M.A, Asgar, Phys. Rev. Letters, 22, 1436 (1969)

25. M.O. Steinitz, L.H. Schwartz, J.A. Marcus, E. Fawcett and W.A. Reed, Phys. Rev. Letters, 23, 979 (1969)

26. G.V. Bunton and S. Weintroub, J. Phys. E, 1, 58 (1968)

27. A.M. Thomson, I.R.E. Transactions on Instrumentation, Vol. 1-7, 245 (1958)

28. G.K. White, Cryogenics, 1, 151 (1961)
For a review of methods of measuring thermal expansion see J.G. Collins and G.K. White, Progress in Low Temperature Physics, edited by C.J. Gorter, Vol. 4 (London: North-Holland), p. 450

29. E. Fawcett, Phys. Rev. B.2, 1604 (1970)

30. V.M. Pudalov and M.S. Khaikin, Cryogenics, 9, 128 (1969)

31. J.S. Kouvel and R.H. Wilson, J. Appl. Phys., 32, 2765 (1961) (Fe)
J.S. Kouvel and C.C. Hartelius, J. Phys. Chem. Solids, 25, 1357 (1964) (Co).
J. Veerman, Thesis, University of Amsterdam, (1964) (Ni)
J. Veerman and G.W. Rathenau, Proc. Int. Conf. Magnetism Nottingham 1961, (London, I.O.P.) p. 737.

32. K.I. Arai, Jap. J. Appl. Phys. (Japan), 11, 1303 (1972)
A.B. Smith and R.V. Jones, J. Appl. Phys. 34, 1283 (1963)
P. Hansen, J. Schuldt and W. Tolksdorf, Phys. Rev. B.8, 4274 (1973)

33. P. Kapitza, Proc. Roy. Soc. A135, 537 (1932)

34. K.L. Dudko, V.V. Eremenko and L.M. Semenenko, Phys. Stat. sol., 43, 471 (1971).

35. B.K. Ponamarev and R.Z. Levitin, Priborg Tekh. Eksp. 3, 188 (1966)

36. J. Ricodeau, D. Melville and E.W. Lee, J. Phys. E., 5, 472 (1972)

Magnetic Resonance

D.J.E. INGRAM
Chelsea College, University of London

1. Introduction

One of the particular features of all the magnetic resonance techniques is that they measure the properties and energy states of individual electrons, atoms, or nuclei, and hence afford a very sensitive probe into the structure and properties of materials in their finest detail. It is, of course, possible to correlate these measurements of the atomic properties and parameters with those that are associated with the bulk material as a whole, such as susceptibility or specific heat; whereas it is often not possible to carry this through in the reverse direction, and calculate the particular atomic energy levels precisely from the properties of the bulk material. The main application of magnetic resonance, is applied to magnetic materials, is therefore to be found in the understanding it can give of the particular energy states of the nuclei, atoms, or electrons, which form the basic components of the material under investigation. Such measurements can also often throw considerable light on the basic interactions within the material, such as exchange, or spin-spin coupling, which are themselves responsible for the main bulk properties of the particular magnetic material being studied.

Although most of the applications of these techniques, in the field of magnetic materials themselves, will in fact be concerned with the interactions between electrons, or atoms, it is nevertheless easiest to introduce the general ideas and principles of the resonance techniques by considering individual electrons, atoms, or nuclei as separate entities first. It is then possible to discuss how the results which would be expected from such isolated systems will be modified by the internal interactions which are present in magnetic materials. At the outset it should probably be made clear that resonance techniques can be divided into two very basic groups, i.e. those associated with the nuclei of atoms and covered by the term 'Nuclear Magnetic Resonance', and on the other hand, those associated with the unpaired electrons in an atom and the energy states occupied by these for which the general term 'Electron Spin Resonance' can be employed.

The basic principle of the resonance method is the same for both of these two cases, as may be shown by considering the free proton and free electron as examples. Both of these elementary particles have a charge and spin associated with them, and even a simple classical picture will associate a magnetic moment with a spinning charge. It is in fact true that both the electron and the proton have an inherent magnetic moment associated with them, although not of the precise magnitude given by the simple classical theory. Thus on the simple classical picture the $\frac{1}{2}.\frac{h}{2\pi}$ units of spin angular momentum associated with the electron should have $\frac{1}{2}.\mu_B$ units of magnetic moment associated with it, where μ_B is known as the Bohr magneton. It is equal to $\frac{eh}{4\pi m}$ and is the amount of magnetic moment that is associated with an electron moving in a circle corresponding to the radius of the first Bohr orbit. It is found in practice that the magnetic moment of the electron spin is about twice as much as it should be on the simple theory, and this is referred to as "the anomalous g-value of the electron spin". A more sophisticated calculation of quantum electrodynamics predicts that the value should actually be 2.0023 times as great as that calculated on the classical theory, the small additional factors coming from the interaction of the electron with its own radiation field.

The same basic idea of a magnetic moment associated with a spinning electric charge can also be applied to the case of the proton, which is found to have a magnetic moment of the opposite sign to that of the electron, when compared with the direction of spin, as would be expected. The magnitude of this magnetic moment is very much less than that associated with the electron, since, as seen from the expression for the Bohr magneton, the mass of the particle comes in the denominator of the term, and hence the very much larger mass of the proton produces a magnetic moment some two thousand times smaller than that of the electron. The particular value of the magnetic moment is again not that precisely predicted by a simple classical picture, and is in fact equal to 5.585 nuclear magnetons, where the nuclear magneton is given by the same expression as for the Bohr magneton, but with the mass of the electron replaced by the mass of the proton.

If we consider completely free electrons, or protons, in a space devoid of any applied magnetic fields, then their spins and associated magnetic moments may point in any direction at random, and all such directions will have the same energy associated with them. If, however, an external magnetic field is applied across such an assembly of free electrons, or protons, then the particles will become sorted into two groups with a definite energy difference between them. Thus, if the case of the electron is taken first, these will not only be characterised by a spin quantum number S, but also by a magnetic quantum number M_S, which defines the resolved component of the spin momentum in the direction of the applied magnetic field. Since successive values of any quantum number must differ by unity, and the maximum value of M_S that can be associated with a spin quantum number of ½ is in fact $M_S = +½$, it follows that in this case there are only two possible allowed states for the electron spins, i.e. those with quantum numbers $M_S = +½$, and those with quantum numbers $M_S = - ½$. The application of the external magnetic field will therefore align the magnetic moments of the electrons either parallel to, or against, the direction of the field itself, and the two groups of electrons so formed will have different energies.

The reason for this energy difference can be seen from a simple comparison with the classical case of small bar magnets placed in the homogeneous field of a large electro-magnet. Such small bar magnets would normally align with their north poles facing the south pole of the electromagnet, and in such a position they would be in stable equilibrium and if displaced slightly would return to their original orientation. Such a position would of course correspond to a minimum of potential energy. On the other hand it would be possible in principle to place these bar magnets in the uniform field with exactly the opposite orientation, so that their north poles were facing the north pole of the electromagnet, but in this case any slight disturbance from the position of exact symmetry would cause them to swing round completely into the more stable position, losing potential energy as they did so. This fact, that the magnetic moments aligned against the field have a higher energy than those lined up with the field, can of course be more simple expressed in terms of simple magneto-statics by the expression for the energy of a magnetic moment lined up in the magnetic field. I.e.

$$\text{Energy} = -\mu.B \cos \theta$$

where μ is the magnetic moment of the dipole concerned, and θ is the angle between its axis and that of B the applied field.

It follows from this brief analysis that the electrons will now be in two groups, with an energy difference between them equal to $g\mu_B B$, as shown in Figure 1(a). The magnetic moment associated with each electron is in fact equal to $\frac{1}{2}g\mu_B B$ the $\frac{1}{2}$ being the value of the M_S quantum number, and the g factor being the anomalous term which allows for the fact that the actual value of the magnetic moment is not the simple value predicted by the classical theory. The Bohr magneton term, μ_B converts the units of angular momentum to those of magnetic moment. It follows that the electrons with their spins aligned against the field are raised in energy by amount of $\frac{1}{2}g\mu_B B$, while those with their spins aligned with the field are lowered in energy by the same amount, to give the total energy difference of $g\mu_B B$. An exactly similar argument can now be applied to the protons, which would all have equal energy whatever the orientation of their spins and magnetic moments in the absence of any applied field, but on application of an external magnetic field they will also be sorted into two groups with an energy difference equal to $g_N \mu_N B$, as indicated in Figure 1(b). In this case the energy difference is in fact about two thousand times smaller than that of the electron case owing to the larger mass, and hence smaller magnetic moment, of the proton, and hence the scale of the two diagrams in Figure 1 is quite different , although the general principle is exactly the same.

Having separated the electrons, or protons, into two energy groups as indicated, then radiation is fed into the system of such a frequency that the energy of its quanta, $h\nu$, is equal to the energy gap between the two groups of particles. Thus, in the case of electron resonance, the resonance condition is fulfilled by bathing the electrons with radiation such that

$$h\nu = g\ \mu_B\ B \qquad 1.1$$

When this resonance condition is fulfilled electrons in the lower energy level can be excited to the high energy level and absorb the incoming radiation as they do so. The actual existence of the resonance phenomena can then be detected by this absorption of the incident radiation. If typical values are put into the resonance equation for the electrons then, for a completely free electron in a magnetic field of 1 tesla (10,000 gauss) it will be seen that the required resonance frequency will be 28,000 MHz, which is a frequency in the microwave, or radar, region. Hence it follows that most of the experimental techniques associated with electron spin resonance will be concerned with microwave electronics, whereas the substitution of typical figures for the nuclear resonance case, and for the proton in particular, would give a resonance frequency of 42.6 MHz for the same field of 1 tesla (10,000 gauss). Hence it follows that the techniques associated with nuclear resonance will be those applicable to more normal radio frequencies instead of the radar techniques of electron resonance.

The accuracy with which present frequencies can be measured is in fact extremely high, and this feature, associated with the other very precise measurements in such frequency ranges, enables the electrons and nuclear resonance techniques to probe the actual energy levels of the electrons, or nuclei, in very precise detail. Since one other chapter of this series is devoted to the application of nuclear magnetic resonance to magnetic materials this technique will not be considered any further in this chapter, but it was felt that a general comparison between it and electron resonance might be helpful at the beginning, to make clear that both of these techniques are based on the same fundamental principles, and the difference of experimental technique arises simply from the much smaller magnetic moments of the nuclei.

It can of course be argued that since the resonance condition has two variables in it, i.e. the magnetic field strength and the frequency of the applied radiation, then it should also be possible to carry out electron resonance at radio frequencies by just reducing the strength of the applied magnetic field. This is in fact quite true, in principle, and some electron resonance measurements have in fact been carried out at radio frequencies of the order of 28 MHz and with applied magnetic fields of 10^{-3} tesla (10 gauss) strength. There is however one very fundamental reason why electron resonance should be performed at as a high a field strength as possible, and this arises from the fact that so far only half of the picture has really been considered. Thus we have ignored the fact that, at the same time as the incoming radiation can cause excitation of the electrons from the ground level to the higher level and become absorbed itself in the process, it can

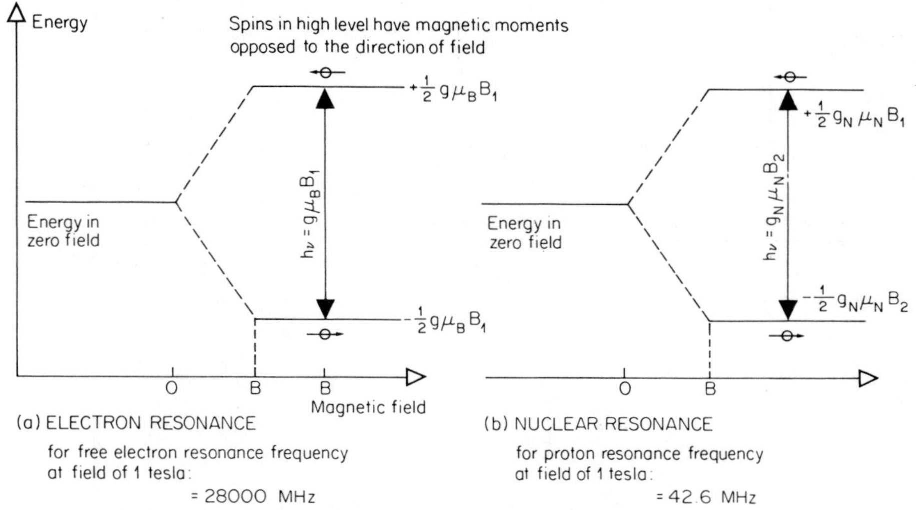

Figure 1 BASIC PRINCIPLE OF MAGNETIC RESONANCE

(a) Energy splitting for electrons in applied magnetic field.

(b) Energy splitting for nuclei in applied magnetic field.

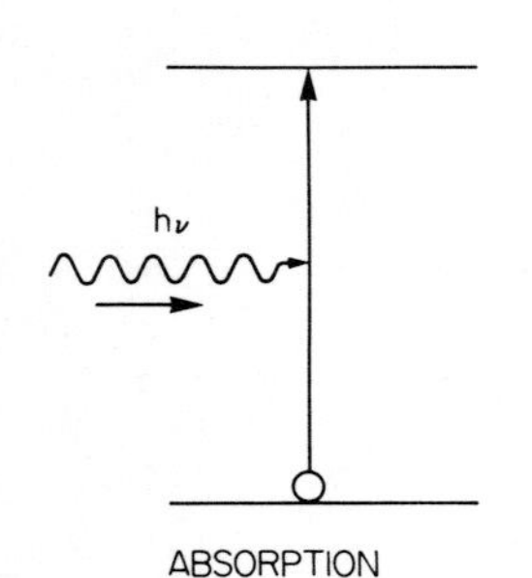

ABSORPTION

Incident quantum raises electron to higher level

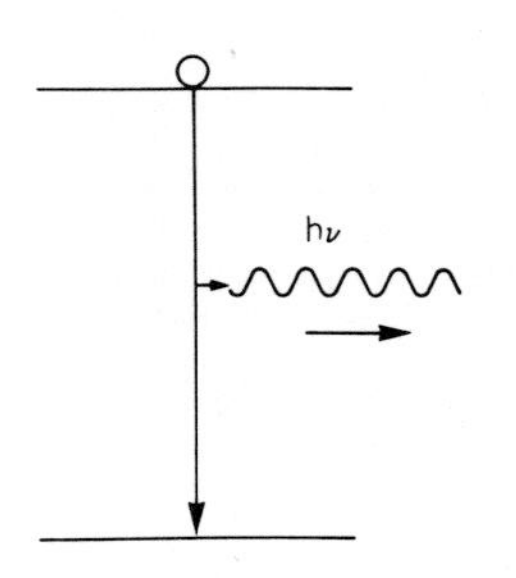

SPONTANEOUS EMMISSION

Electron reverts to ground state - emitting quantum in process

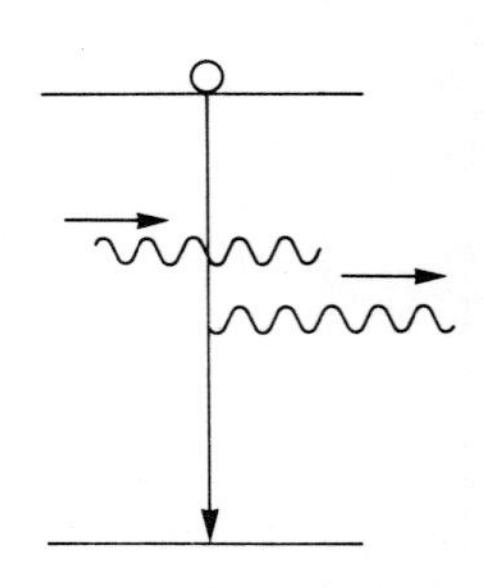

STIMULATED EMMISSION

Incident quantum stimulates electron already in higher level to emit second quantum - in phase with first

(a)

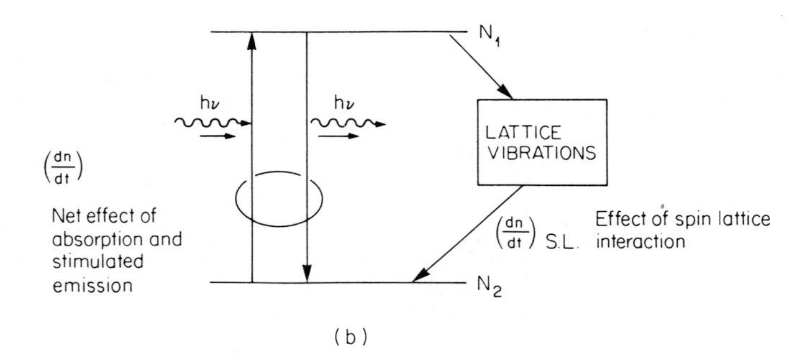

Figure 2 — ENERGY LEVEL POPULATION

(a) The different types of interaction with radiation, absorption transfers electrons from the ground state to the higher level, whereas both spontaneous and stimulated emission do the reverse.

(b) Energy level population when in thermal equilibrium. The net reduction in $(N_2 - N_1)$ produced by the radiation is balanced by the effect of the spin-lattice interaction.

alternatively produce stimulated emission by causing electrons already in the higher level to revert to the ground state, as shown in Figure 2 (a). In his early theory of radiation processes Einstein[1] showed that the coefficients for absorption and stimulated emission were in fact equal, so that if there were the same number of electrons in the ground state and in the excited state, then the incoming radiation would cause just as many transitions from the lower state to the higher state, as it did from the higher to the lower state. Thus the absorption of radiation would be exactly counterbalanced by an equal additional amount of stimulated emission and hence there would be no net resultant observed change in the radiation level. The only reason why there is in fact a net absorption which can be observed, is that there are fewer electrons in the higher energy level and thus there will, in practice, be somewhat more absorption than there is stimulated emission. The magnitude of this net absorption will depend crucially however on the difference in the energy level populations, since this is normally given by a Maxwell-Boltzman distribution, such that

$$\frac{N_1}{N_2} = e^{-g\mu_B B/kT} \qquad 1.2$$

Since it will be much easier to detect the electron resonance if the net absorption is large, it follows that the fraction $\left(\frac{g\mu_B B}{kT}\right)$ should be made as large as possible, and this is the reason why electron resonance is carried out at as high a field strength as possible, and often at as low a temperature as possible.

The distribution of electrons between the two energy levels is shown schematically in Figure 2(b), together with the effect of an additional interaction known as 'spin-lattice interaction' which is necessary if the unequal population distribution is to be maintained. This interaction, together with its effect on the width of the energy levels and absorption lines is considered further in section 4.2.1.

In this section we have been considering completely free electrons, and the energies of their spins and magnetic moments in an applied magnetic field. In the next section this consideration will be extended somewhat to include the three different cases of magnetism, as they arise in real material, and then the following section will take up further detailed consideration of the actual experimental techniques that are required to investigate these specimens in practice.

2. Different Types of Magnetic Resonance

In the same way that the magnetism of different materials can be divided into three basic types, i.e. diamagnetism, paramagnetism and ferromagnetism, so the kinds of resonance which can be observed in different material can be classified into these groups. Thus diamagnetism of all materials is caused by a change in the orbital motion of the electrons and is not connected at all with the magnetic moment associated with the electron spin. The analysis of the preceding section would therefore not apply to diamagnetic material, where there are no unpaired electron spins, but it is nevertheless sometimes possible to observe another type of resonance which is associated with the orbital motion of free electrons and not due to their spin magnetic moment at all. This type of resonance is termed 'Cyclotron Resonance' and is normally only observed for the charged carriers in metals or semiconducting material.

The basic mechanism of cyclotron resonance is in fact an interaction of the charge on the electron with the electric field component of the incident microwave radiation, and as such it is not really a "magnetic resonance"at all. On the other hand it does require the application of a large external magnetic field, and since the experimental equipment required for its observation is almost identical to that of electron spin resonance they are often considered together. It is, however, not really of very great relevance to studies on magnetic materials, but a brief description of the basic principle is given later in this section so that the different types of resonance can be compared.

In contrast to cyclotron resonance, the analysis for the last section applies very directly to all paramagnetic types of material, where there are unpaired electrons possessing magnetic moments which interact directly with the externally applied magnetic field. If these electrons can be considered as individual entities, with no strong interactions between them, then they can be considered as relatively free; the material is in a paramagnetic rather than a ferromagnetic state, and a normal paramagnetic electron spin resonance signal will be obtained from it. In fact the majority of specimens that have been studied by electron spin resonance have been these paramagnetic types of material, and the information that can be deduced from such studies will be summarised in detail in later sections.

If the interaction between the unpaired electrons inside the material is very strong, this will of course produce a ferromagnetic, anti-ferromagnetic, or ferrimagnetic system, and all these can also be studied by resonance methods. The crucial difference here, compared with the straightforward case of paramagnetic resonance, is that the unpaired electrons not only experience the externally applied magnetic field but also the internal field which is produced by the strong interaction, within the specimen. The condition for resonance is thus very considerably shifted away from that predicted by the effect of the external field alone, and one of the applications of such ferromagnetic, or anti-ferromagnetic, studies is a direct internal measurement of the strength of this internal field. Although the basic principles are still the same as those outlined in section 1, the practical considerations which result from the presence of these large internal fields necessitate a separate consideration for ferromagnetic resonance, and this will be given towards the end of this chapter.

2.1 Diamagnetic or cyclotron resonance

The principle behind the cyclotron resonance is exactly the same as that of the original cyclotron, which was designed to feed power from an oscillating r.f. electric field to orbiting electrons and thus build up their energies. In the case of cyclotron resonance applied to solid state specimens the r.f. electric field is provided by the incoming microwave radiation; the electrons, which are free to move within the material, are made to move in circular orbits by applying an external magnetic field at right angles to the oscillating electric field as is shown in Figure 3 (a). It can be seen from this that, if the orbiting electron is moving in such a direction as to be in phase with the driving microwave electric field, then it can be accelerated on each side of its orbit and thus continuously take energy from the field as it rotates. In contrast to this an electron which is initially moving in anti-phase to the applied electric field will be decelerated by the field, but in the process it will be moved out of the interaction area. Hence there will be a net absorption of radiation by the orbiting electrons. Although in principle, cyclotron resonance should be observed from any material containing free electrons, or other charge carriers, in practice the conditions for its observation are extremely severe. Thus when the question of the width of the absorption lines, and the broadening processes which give rise to these, are considered in detail it is found that it is only likely that such absorption will be detected from extremely pure materials, since otherwise rapid scattering of the orbiting electrons will destory their phase coherence with the incident microwave electric field before any significant number of orbits have been completed. In fact in the early studies of cyclotron resonance it was only possible to observe such absorptions from the electrons, or holes, in a limited number of semiconductor materials, such as germanium and silicon, and for a long time no observation at all could be obtained from electrons in metals.

This lack of observations from metals themselves was due to the fact that the very large numbers of the electrons in metal conductors produced a severe bunching of the electrons in the orbits around the phase position of maximum interaction. This large bunching of electrons itself then defocused the electron interaction and hence broadened the absorption line beyond the point of detection . It appeared at first that there was very little that could be done about this particular phenomenon, since it was produced by the inherently larger number of charge carriers within the metallic conductors. Some years after the observation of cyclotron resonance in semiconducting material, however, it was suggested that it might be possible to observe it from metallic specimens as well if the applied field configuration was somewhat altered to that shown in Figure 3 (b) . Here the applied magnetic field is shown parallel to the surface of the metallic specimen, and also parallel to the direction of the oscillating electric field. It will be seen that the orbiting electrons will now be moving in orbits which take them close to the surface of the metal and then away from it, and in the process they will be coming into, and going out of, the skin depth of the microwave electric field in the metal itself. Thus good conductors will have a relatively small

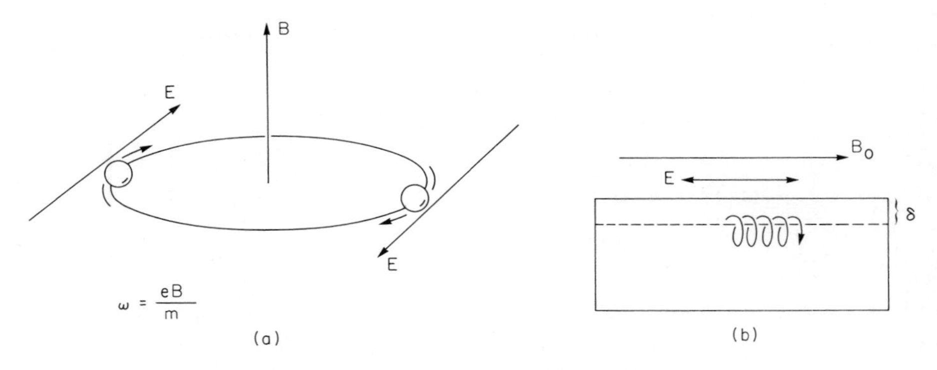

Figure 3 BASIC PRINCIPLE OF CYCLOTRON RESONANCE

(a) Normal configuration of applied fields.

(b) Configuration suggested for good conductors so that the interacting electrons can spiral beneath the 'skin-depth'.

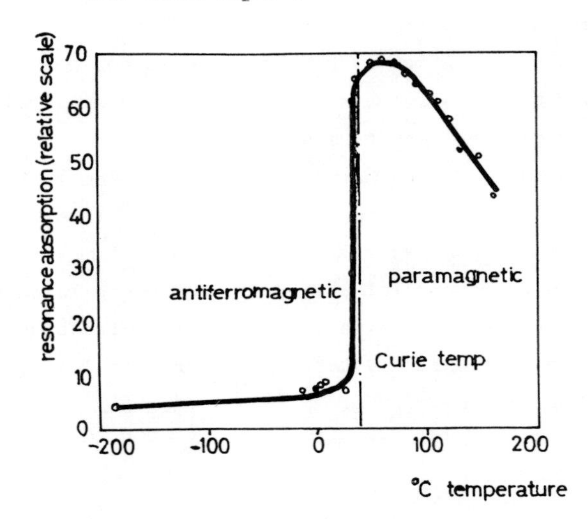

Figure 4 ANTI-FERROMAGNETIC RESONANCE IN Cr_2O_3

The rapid disappearance of the normal paramagnetic resonance signal as the temperature is lowered through the Curie point can be clearly seen.

skin depth, and below this any electrons will effectively be shielded from the electric field. A situation has now been achieved where the orbiting electron will only interact with the microwave electric field when it is at the top of its orbit within the skin depth of the material. In the case shown it will be seen that the particular electron will now be given a small increment of added momentum to the right-hand side, each time it comes up within the skin depth. It can thus absorb energy from the incident radiation field, by continuing to move to the right, and it will in fact be shielded from the electric field when this has the opposite direction. The advantage of this arrangement is that each individual electron can now be moved along steadily in the direction of the electric field, and no phase bunching of the electrons will occur. As a result the defocusing and broadening effect on the actual absorption will also be removed and hence a simple absorption line, corresponding to the interaction of the electric field with the electrons at the top of their orbits, can be observed. This possibility was tried out experimentally very soon after it had been suggested, and cyclotron resonance has now been observed from a large number of metals, as well as the semiconductors earlier studied.

The detailed information that can be deduced from such cyclotron resonance measurements will not be considered here since it is not really appropriate to the specific study of magnetic material, but it will be evident that the crucial parameter that comes immediately from such observation is the "effective mass" of either the electron or the hole. Thus the condition for cyclotron resonance is identical to that of the cyclotron itself, i.e.

$$\omega = \frac{eB}{m} \tag{1.3}$$

and for an electron, or hole, moving within the potential wells of solid material the mass that is observed is the "effective mass" of the charge carrier. The technique of cyclotron resonance thus allows a very direct measurement of this important parameter, and if more than one charge carrier is present at a time each of these can be measured, and their anisotropy determined as well. Further information on the internal interactions within the specimen can also be deduced from the linewidth of the observed resonances and the general analysis of this is somewhat similar to that given later when the linewidths of the electron resonance signals themselves are discussed.

2.2 Paramagnetic resonance - E.S.R.

As explained earlier, the vast majority of studies on magnetic resonance of electrons have been concerned with paramagnetic materials to which the simple analysis of section 1 applies directly. Such studies of electron paramagnetic material are normally now included under the general heading of electron spin resonance, although of course, ferromagnetic resonance is basically concerned with electron spins as well. In the case of the paramagnetics however the interactions from the surrounding electron spins can be ignored to a first order and the absorption can be treated as due to the interaction of the radiation field with the individual electron. Internal interactions can then be treated as perturbations on the main energy interaction. This method of treatment is developed in sections 4.1 - 4.5, where the different parameters which can be associated with the electron resonance absorption are considered in more detail. It may be noted here however that it is often possible to obtain very precise measurements on such parameters as the exchange interaction energy from superhyperfine splittings of the electron resonance spectrum and such information can often be of very direct relevance to studies on ferromagnetic material itself. Another useful feature of the paramagnetic studies is that material can often be diluted by isomorphous dimagnetic compounds and hence exchange interactions can be reduced in strength so that they do not obliterate the features of the spectrum. In this way studies on paramagnetic systems can often be used as models of ferromagnetic systems, and the two can be usefully correlated in this way.

2.3 Ferromagnetic resonance

As mentioned earlier, ferromagnetic resonance is basically the same phenomenon as paramagnetic resonance except that the electrons being studied are experiencing the strong internal field of the ferromagnetic material as well as the applied external field. The strength of the combined fields will depend very much on the anisotropy constants of the material itself, together with any demagnetising factors associated with the specimen, although in the simplest cases the value of the external field B_o is replaced by $(B_m B_o)^{\frac{1}{2}}$ in the resonance condition, where B_m is the induction in the ferromagnetic material.

$$B_m = B_o + \mu_o M \qquad \left[B_m = H + 4\pi M\right]_{emu}$$

The actual experimental observation of ferromagnetic resonance was first made by Griffiths[2] in 1946 when the variation of the Q of the microwave cavity was measured. This had ferromagnetic material electroplated on to one end wall and the variable magnetic field was applied across it. The initial theory of this resonance phenomena was undertaken by Kittel,[3] who derived expressions to account for both the internal magnetisation and the effect of the demagnetising fields, and the details of this theory will be taken up in the later sections. One of the general difficulties encountered with ferromagnetic resonance is the rather large width of the absorption lines which are obtained, which often obscures the fine structure or the hyperfine structure which can be observed in paramagnetic resonance itself.

In the same way as ferromagnetic material can be studied so antiferromagnetic resonance can also be observed from antiferromagnetic lattices, and one of the striking results obtained from these measurements is the sudden broadening and disappearance of the normal paramagnetic absorption line as the Curie point is approached and passed. This was first noticed in Cr_2O_3 by Maxwell and McGuire[4] and the results are illustrated in Figure 4. where it is seen that the signal intensity falls sharply just below the Curie point at 311^oK. Similar results are also obtained when many of the antiferromagnetic manganese inorganic compounds are studied, and in each case the disappearance of resonance is, of course, due to the fact that the anti-ferromagnetic interactions have set in just below the Curie point, and shifted the resonance condition to a range of applied field and frequencies well beyond that normally employed. The new conditions for resonance can however then be picked up below the Curie point and precise information on the value of the internal magnetic fields can then be obtained. Examples of such studies will also be summarised in section 5.

2.4 Experimental requirements

It will be seen that all three of these basic types of resonance require very similar equipment, i.e. a source of microwaves of the appropriate frequency together with an electromagnet to produce the appropriate magnetic field. In both paramagnetic and ferromagnetic resonance the incoming microwave radiation field interacts with the magnetic moment of the electron, via the magnetic vector of the microwaves, and hence the specimens themselves need to be placed in such a position that they are in a region of the maximum of the microwave magnetic field strength. This can be achieved by appropriate design of the cavity resonator system as discussed in section 3.1. On the other hand it has already been seen that in cyclotron resonance the interaction is between the microwave electric field and the electric charge of the electron , and hence in this case the specimen needs to be placed in the region of maximum microwave electric, rather than magnetic,field strength. Apart from this difference however the techniques for cyclotron resonance are basically the same as those for electron resonance and will not be discussed in any further detail. The only additional point that should be made in this connection however is that the strength of the electric dipole coupling is many orders of magnitude greater than that of the magnetic dipole coupling, and hence much smaller concentrations of electrons, or other charge carriers, can be detected by cyclotron resonance than for the corresponding magnetic case. This inherently greater sensitivity is somewhat offset however by the much larger linewidths that are normally observed in cyclotron resonance studies, but the whole question of sensitivity is discussed in much more detail in section 4.1.

3. Experimental Techniques

3.1 The basic electron resonance spectrometer

It will be evident from the previous discussion that there are two main requirements on the experimental side for any electron resonance spectrometer, i.e. (i) a large uniform magnetic field, (ii) a stabilised microwave frequency source in the correct wavelength region. The general way in which these can be assembled to form an actual spectrometer is illustrated schematically in Figure 5 where the basic

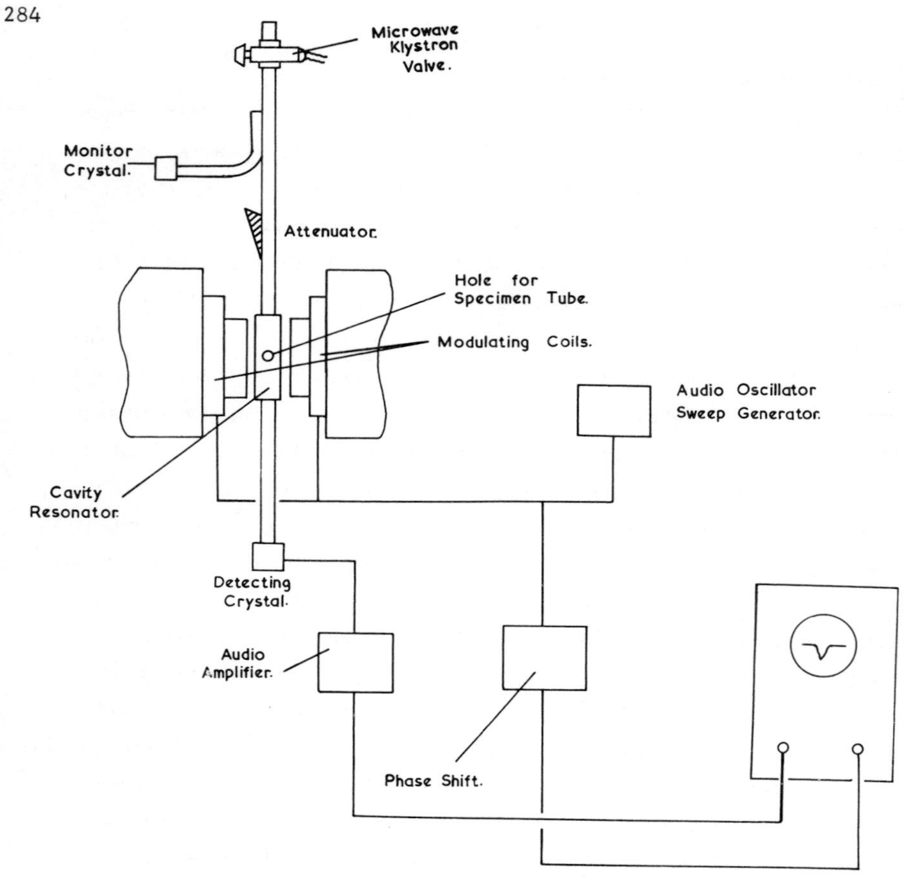

Figure 5 SIMPLE 'CRYSTAL-VIDEO' ELECTRON RESONANCE SPECTROMETER

items are, first the microwave radiation source, secondly the absorption cell in which the specimen is inserted, thirdly the large uniform magnetic field into which the absorption cell containing the specimen is itself inserted, and then finally some means of detecting the level of the microwave radiation in the absorption cell and determining whether any absorption has in fact taken place. These items are, of course, the essential features of any spectrometer working at any wavelength. It is noticeable, however, that in this particular case the dispersive element, corresponding to a prism or grating can, in fact, be dispensed with, because the microwave source, which produces the electromagnetic radiation, is so precisely built and designed that extremely monochromatic radiation can be obtained from it.

This single frequency is then fed down a waveguide to the absorption cell which takes the form of a resonant cavity, as shown in Figure 5. The cavity resonator, which is adjusted to be of such a length that it is on tune for the incoming microwaves, serves to concentrate the radiation in its standing wave pattern and the sample is placed in it, in a position of maximum microwave magnetic field. The cavity itself is held between the pole faces of a strong electromagnet, and provision is often made to cool the cavity and specimen by immersing them in a dewar containing liquid nitrogen, hydrogen or helium. The level of power in the cavity can then measured by tapping a fraction of this out along the output wavelength to a crystal detector, where the microwave radiation is converted into a d.c. signal. If a small a.c. magnetic field modulation is also applied at the same time, the field value can be swept through resonance twice in each cycle, and hence the output signal from the crystal will be modulated at twice the frequency of the field modulation. In this way a.c. techniques can be used to amplify and display the absorption line on the screen of an oscilloscope as indicated in Figure 5. A simple spectrometer of this type is often referred to as a "crystal video spectrometer" since it employs simple a.c. modulation of the magnetic field and displays the absorption directly on an oscilloscope screen. It contains all the basic essentials of an electron resonance spectrometer but has rather a low sensitivity, due to the very high additional noise that is present in silicon crystal rectifers at the low frequencies used for the modulation. The ways in which this noise can be reduced, and hence more sensitive spectrometers can be built in practice, are now summarised.

3.2 Microwave bridge systems

The main limitations and drawbacks of the simple crystal-video spectrometer system are due to the properties of the detecting crystal. There are two particular properties of these silicon, or germanium, rectifying crystals that necessitate basic design differences in high sensitivity spectrometers. The first of these properties is known as the conversion loss , and it effectively measures the efficiency of the crystal in converting the incoming microwave radiation to d.c., or low frequency modulation. This conversion factor varies noticeably with the actual level of the microwave power falling on the crystal and hence, with the mean d.c. current flowing through the crystal. This variation is shown in Figure 6, and it can be seen that the conversion loss (and thus, the inefficiency of detection) is very high at low values of input microwave power and mean crystal current, but falls rapidly as these increase. It flattens to quite a low value once the current has risen above about 1 milliamp. Consideration of this parameter by itself, would, therefore, suggest that the crystal detector should always be operated with a fairly large amount of microwave power falling on to it, so that the mean detecting current was well above 1 milliamp.

There is, however, one other property of the crystal which must be considered at the same time. This is the excess noise which the crystal itself produces, over and above the normal random thermal noise of its equivalent resistance. It will be seen later that this noise varies very rapidly with the frequency of detection, but it also varies quite markedly with the amplitude of the mean detected current flowing through the crystal. This variation of noise against detected current is also plotted on Figure 6, and it can be seen that it rises linearly with increasing magnitude of the current flowing through the crystal, and hence, from the point of view of the reduction of excess noise, the mean crystal current should be kept as small as possible. As a result of both of these effects together there is an optimum

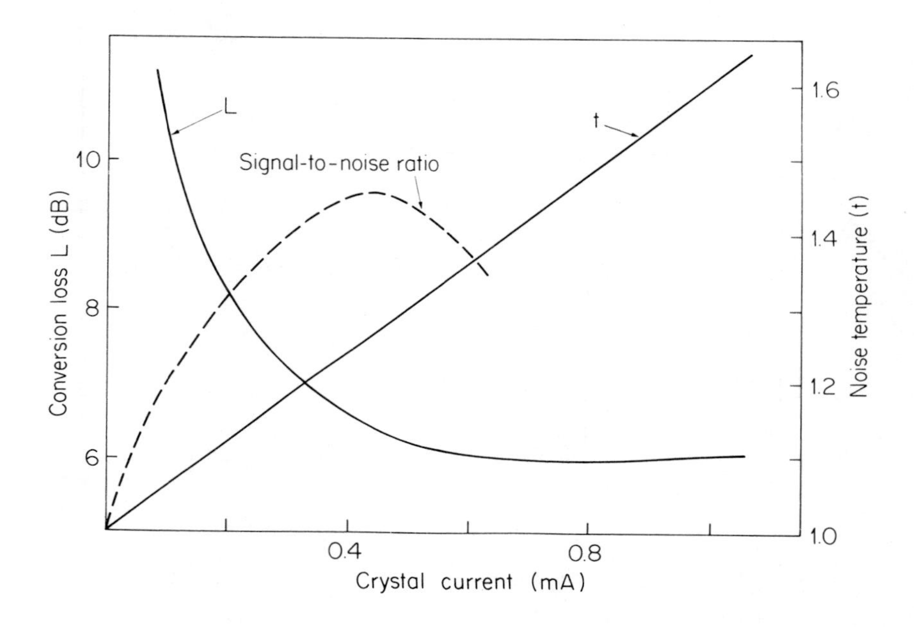

Figure 6 VARIATION OF DETECTOR CRYSTAL CONVERSION LOSS AND EXCESS NOISE WITH DETECTED CURRENT

It is seen that there will be a broad optimum in the resultant signal-to-noise ratio at about ½ milliamp detector current.

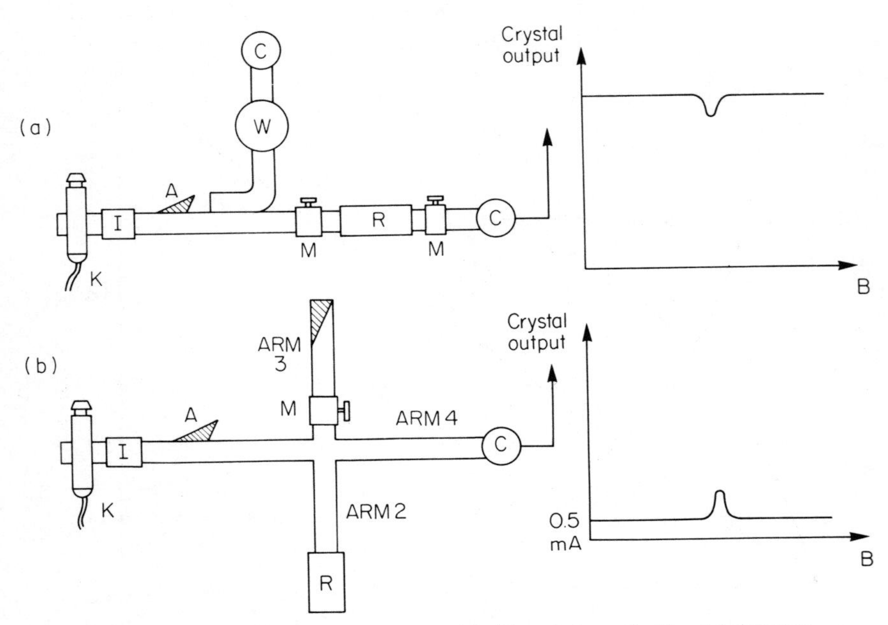

K = Klystron I = Isolator A = Attenuator W = Wavemeter C = Crystal detector
R = Cavity resonator M = Matching unit

Figure 7 POWER BALANCING BY MICROWAVE BRIDGE

a) Simple transmission system producing small change in large output signal.

b) Bridge system which balances out most of power to produce signal in relatively small total output.

crystal current at about ½ milliamp, where the conversion loss has fallen to a reasonably low value, and where the excess noise has not become too large. This is illustrated in Figure 6 as a broad maximum in the overall signal/noise ratio. It is for these reasons that crystal detectors are normally operated with a crystal current of about ½ milliamp through them. It will be seen later that the sensitivity of all the electron resonance spectrometers is limited by the noise produced in these crystal detectors, and it is therefore very important to operate these under optimum conditions.

One of the major disadvantages of the simple transmission system illustrated in Figure 5 now becomes apparent. In order to obtain a reasonable absorption signal from the specimen itself, the microwave power in the cavity should be made as large as possible. It can also be shown that the coupling out from the cavity to the waveguide will have an optimum value, and hence, a certain fraction of the power concentrated in the cavity will be coupled out to the detecting crystal. The power level in the cavity and the power falling in the detecting crystal are, therefore, not independent of one another, and increasing the power level in the cavity to produce a large amount of absorption may well reduce the ultimate sensitivity, by creating too much excess noise in the detecting crystal itself. Any spectrometer which aims at high sensitivity must, therefore, have some means of allowing independent adjustment of the microwave power in the cavity and of the microwave level at the detecting crystal. There are, in fact, two ways in which this may be done.

The most straightforward way of obtaining independent adjustment of the crystal current is to use some form of microwave bridge system, instead of the simple transmission system of Figure 5. The basic principles of such a bridge system is illustrated in Figure 7(b), and its operation can be contrasted with that of the simple transmission system as represented schematically in Figure 7(a). In most practical applications the electron resonance spectrometer is being used to measure small absorption signals, due to small concentrations of unpaired electrons, and thus the actual microwave absorption will only be a small percentage of the total microwave power passing through the cavity. If a transmission system is used, as illustrated in Figure 7(a), this will then be detected as a small dip in a large steady d.c. output current, as shown, and it is evident that the crystal will then be operating in a region of high excess noise. To avoid this, the microwave cavity can instead be placed in the arm of a microwave bridge, so that, when the bridge is balanced, the power from the the klystron is fed equally to the microwave cavity and the third arm of the bridge, containing a matched load as shown, and no power then passes on to the fourth arm containing the detecting crystal. When resonance absorption takes place the bridge then becomes unbalanced and a resultant signal is passed on to the fourth arm and detected by the crystal. If the microwave bridge were completely balanced in the "off-resonance" position no signal would appear at the detecting crystal and the crystal would, therefore, be working in a region of high conversion loss. The crystal can be adjusted, however, to work at its optimum value by slightly unbalancing the bridge before resonance occurs, so that a standing crystal current of about ½ milliamp is produced in the crystal, which will then be operating under optimum conditions. The resonance absorption will now occur as a relatively large change in the detected microwave power as indicated on the right hand side of Figure 7(b).

This independence of crystal current, from the power level in the microwave cavity, is an essential feature of any modern electron resonance spectrometer and hence they must all employ some form of microwave bridge, or microwave bucking system. The actual microwave components which are used in such a bridge, vary somewhat. Until recently these were normally composed of directional couplers, or "magic T's", but during the past few years the advent of non-reciprocal microwave circulators have made several improved designs a possibility.

3.3 Other sources of crystal noise

The adjustment of the d.c. level of the crystal to an optimum is not the only way in which its noise can be minimised however. The excess noise of the crystal varies with the frequency of modulation as well, and there is in fact an inverse relation between it and the frequency over a very large range, as illustrated in Figure 8.

In this figure, the excess noise produced by the detecting system is plotted against the frequency of modulation used in the detection. It will be seen that there is a straight line graph on the left-hand side, which falls from a large value at the low frequency end, towards the zero axis at higher frequencies . This represents the excess flicker noise due to the detecting crystal itself, and this graph explains why a simple crystal video spectrometer of the type shown in Figure 5 will be inherently very noisy. It follows that high sensitivity in which negligible noise is added by the detecting system, will only be obtained if modulation frequencies of much higher value than the audio range are employed. This fact by itself would suggest that it is only necessary to work at as high a frequency of modulation as possible. This is not true, however, since above about 50 MHz the noise of the I.F. amplifier becomes appreciable, and this is represented by the curve which rises on the right-hand side of Figure 8. It follows that if these two curves are added together to give the total excess noise contributed by the detecting system, then a broad minimum is obtained centred on about 30 MHz as shown in the figure. This suggests that the maximum sensitivity in any microwave detecting system will, therefore, be obtained by employing an intermediate frequency of about 30 MHz, and it is of course for this reason that radar sets use intermediate frequencies of about this value. One of the possible methods of obtaining high sensitivity is to use a superheterodyne system similar to that employed in radar, and with it an I.F. frequency of about 30 MHz. This produces a very complex piece of equipment, however, and if similar sensitivity can be obtained without the addition of the local oscillator klystron and other ancillary apparatus a large number of difficulties can be avoided.

It is found in practice that the excess noise contributed by the crystal at 100 KHz falls to a very small amount and hence modulation and detection at these frequencies is almost as good as that at 30 MHz. It is relatively simple to design a spectrometer which employs 100 KHz magnetic field modulation, and this is, in fact, the system employed by most of the commercial spectrometers now on the market. The 100 KHz magnetic field modulation can be applied to the sample either by a simple loop or wire, wound around the sample and placed inside the cavity resonator, or, alternatively, by embedding a small pair of modulating coils in the wall of the resonant cavity. In the second case the walls of the cavity must be sufficiently thin to let the 100 KHz modulation through without attenuation, but sufficiently thick to act as a short circuit for the microwave frequencies. This can be readily achieved in practice by either coating the inner wall of a ceramic cavity with a silver lining, or casting an araldite mould around an electro-formed cavity wall. In both cases the thickness of the wall is adjusted to be greater than the skin depth of the microwave frequency but less than the skin depth at 100 KHz. Examples of both of these forms of applying the high frequency modulation are shown in Figure 9. In the one case the simple single loop is inserted inside the reactangular cavity and is shown located in the centre, while, in the other case, the Helmholtz coil system, which is placed around the thin walled cylindrical cavity, is indicated in an expanded form.

3.4 Spectrometer systems

The basic requirements for the design of high sensitivity electron resonance spectrometers are now clear. In the first place the simple transmission system must be replaced by some kind of balanced bridge, or similar microwave system, and in the second place, the actual frequency of modulation or detection must be well up in the hundreds of kilocycles, or megacycle region. These requirements can be brought together in two basic forms of spectrometer. The first is called a "High Frequency Modulation Spectrometer" and normally employs a magnetic field modulation of 100 KHz, and thus obtains high sensitivity, without the complexity of an extra klystron. The second basic system employs a superheterodyne system, with a second microwave source to produce an intermediate beat frequency in the 30 MHz region - in a slightly better frequency range, so far as its excess noise is concerned. The sensitivity of such superheterodyne spectrometers is not normally very much greater than that of the much simpler 100 KHz system.

The superheterodyne spectrometers do have one great advantage, however, and that is that they do not produce any additional line broadening in the spectra. Thus any modulation of the resonance condition itself, whether it be in the form of modulation of the input microwave frequency, or of the magnetic field, will produce sidebands on the absorption lines, the frequency deviation of which is equal to

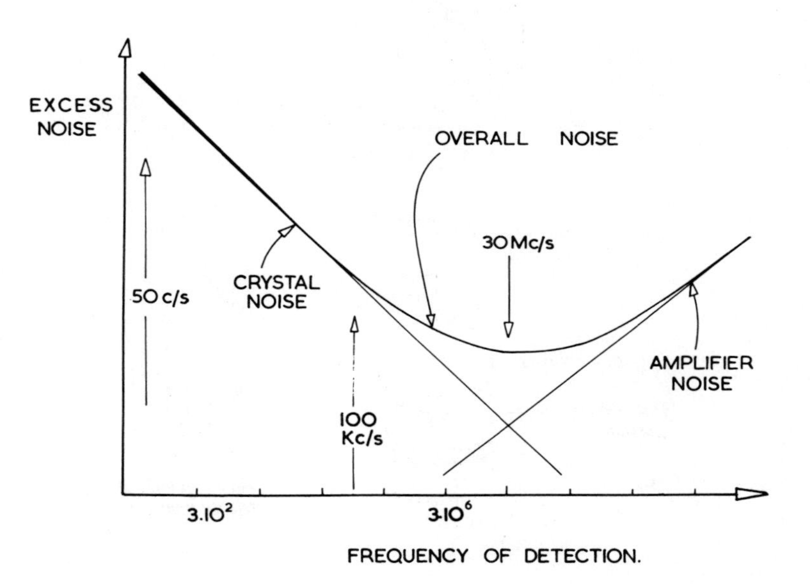

Figure 8 VARIATION OF EXCESS CRYSTAL NOISE WITH FREQUENCY OF DETECTION

The linear variation of crystal noise with f is shown on the left-hand side, while a typical rising line, due to the intermediate frequency amplifier, is shown on the right.

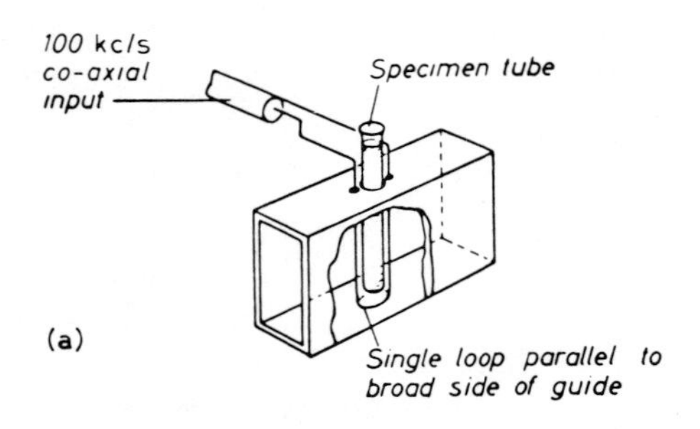

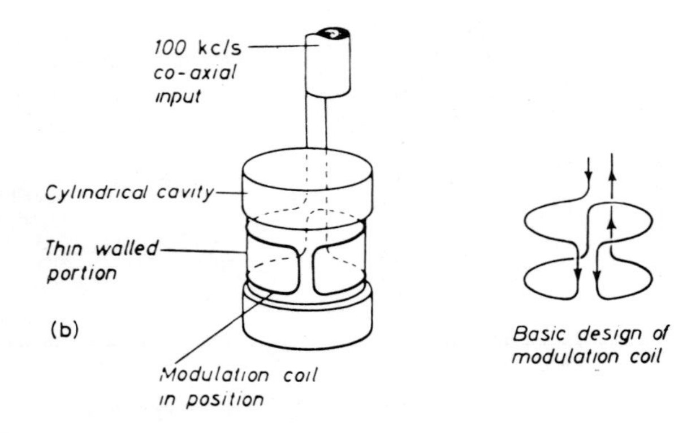

Figure 9 METHODS OF PRODUCING 100 KHz MAGNETIC FIELD MODULATION

a) Internal coil within rectangular cavity.

b) External Helmholtz coils around thin-walled cavity.

the frequency of the modulation. Thus frequency modulation at 100 KHz will produce sidebands on the absorption lines spaced at about 30 milligauss separation. If broad absorption lines from solid state specimens are being studied such additional broadening will not be noticeable, but if free radicals are being studied in solution, or defects in non-proton containing solids, an additional broadening of this amount might well obscure hyperfine splittings. Hence, if both high resolution and high sensitivity are required at the same time, as is often so in free radical or irradiation damage studies, the superheterodyne spectrometer systems may be a necessity. Since most of the applications to magnetic materials are concerned with wide absorption lines, rather than very fine splittings, only the 100 KHz magnetic field modulation spectrometer will be considered in any more detail, and more information on the superheterodyne spectrometers may be found in standard books on E.S.R. if required.[5,6]

3.5 High frequency field modulation spectrometer

A block diagram of a typical 100 KHz magnetic field modulation spectrometer is shown in Figure 10 and this represents the basic design of several commercial spectrometers currently available. It will be seen that this spectrometer employs a microwave bridge system in place of the simple transmission type cavity, previously discussed, although some of the more modern spectrometers now use microwave circulators instead of the older bridge elements since these allow better transfer of microwave power through the system.

The 100 KHz signal itself originates from the master oscillator shown in the centre of the diagram and is fed through a power amplifier to the modulating coils around the sample as indicated. The impedance of the output stage of this power amplifier needs to be matched to the impedance of the modulating coils which is normally very low, hence large high frequency currents of low voltage are produced in the modulating system. Even so the strength of the actual magnetic field produced at this frequency at the site of the specimen is normally relatively small and hence the magnetic field modulation is now used to sample the gradient of the absorption line rather than sweep through it, as in the previous case of the simple crystal video system. The way in which this 100 KHz modulation detects and reproduces the electron resonance absorption is in fact indicated diagrammatically in Figure 11. It is clear from this figure that the actual signal obtained from the spectrometer is proportional to the first derivative of the absorption line contour, and that the magnitude of the modulation sweep must be kept below half the line width of the absorption if distortion is to be avoided.

The 100 KHz modulation of microwave signal is detected by the crystal at the end of the microwave run and passed on to a narrow band amplifier, after which it is mixed in a phase sensitive detector, so that the noise of the previous amplifier stages can be effectively eliminated, and thus the sensitivity of the detection and display system depends only on the band-width of the actual recording equipment.

As well as this detection and display system most spectrometers also employ some kind of automatic frequency-locking circuits to maintain the frequency of the klystron at the correct value. There are essentially two different methods whereby this may be done. The frequency of the klystron can either be locked in absolute terms and in this case comparison with the harmonics of a quartz crystal standard, or the frequency of a high Q cavity resonator, can be made. It is then relatively simple to devise a correcting circuit so that any drift of the klystron frequency from these external references will produce an error voltage which can be fed back to the klystron reflector itself and made to counteract the drift that is produced. This method has considerable advantages if absolute 'g' values are to be determined, but on the other hand has the great disadvantage that the klystron will not follow any drift in the resonance frequency of the cavity containing the specimen and distortion of the output will then result.

The alternative method of automatic frequency control is to lock the frequency of the klystron to the cavity in which the specimen itself is placed. In this way it is possible to ensure that the incoming microwave signal to the cavity resonator is always at the correct resonance frequency and hence ensure the microwave magnet field strength is concentrated at the site of the sample and that no distortion of the line shape will occur. On the other hand the absolute magnitude of the frequency is no longer determined by any external reference, and hence this method of frequency locking is not

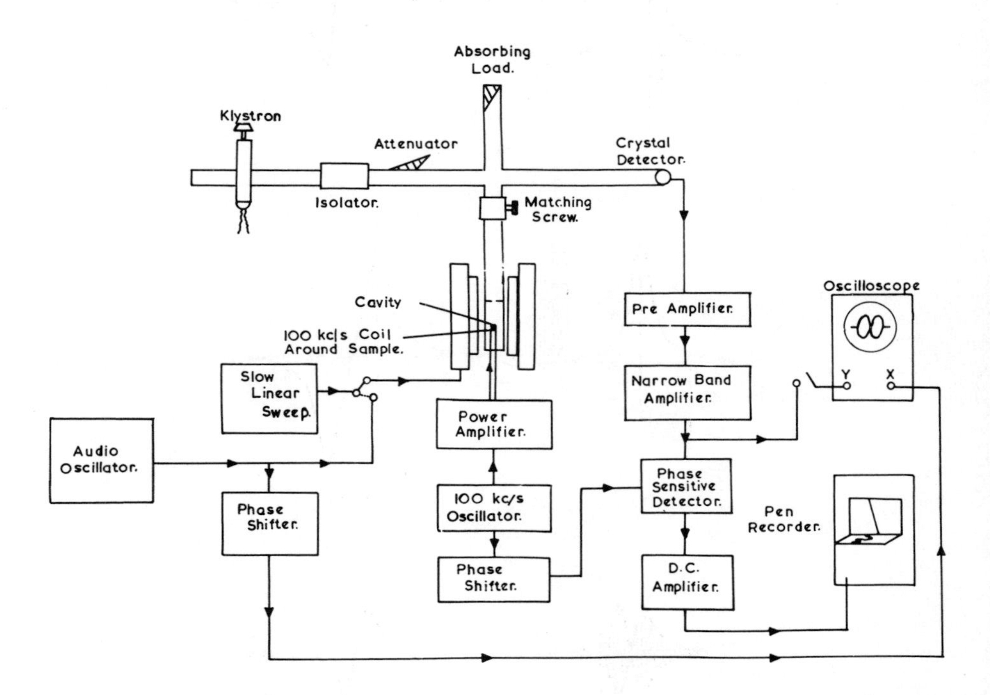

Figure 10 BLOCK DIAGRAM OF TYPICAL 100 KHz MAGNETIC FIELD MODULATION SPECTROMETER

The modulation is provided by the central 100 KHz oscillator, which also feeds a reference signal to the phase-sensitive detector.

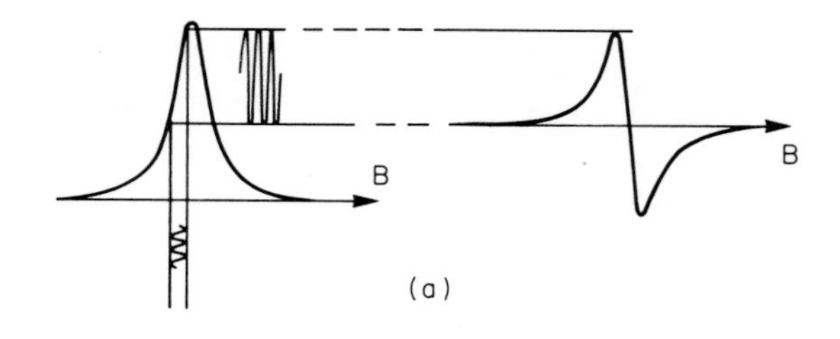

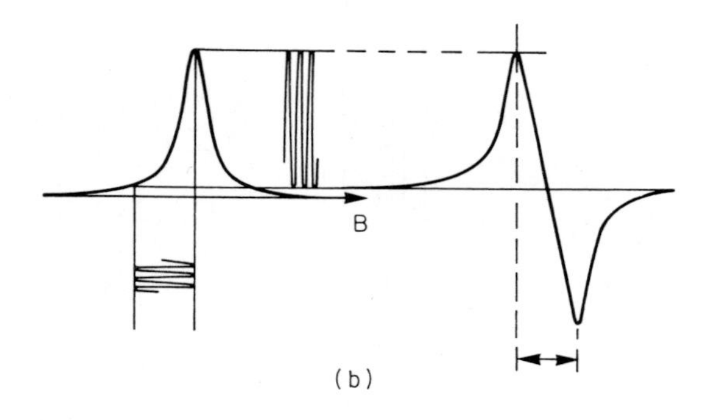

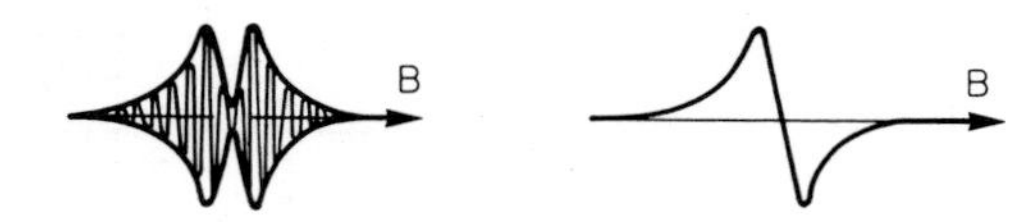

Figure 11 PRINCIPLE OF 100 KHz MODULATION

The way in which the first derivative output is produced by field modulations of different magnitudes is shown for

a) under half the linewidth,

b) twice the linewidths,

c) shows the first derivative before phase-sensitive detection, and

d) the pen recorder trace after phase-sensitive detection.

advisable if the greatest accuracy in 'g' value determination is required. The basic features of either of these automatic frequency control circuits can very easily be added to the diagram of Figure 10 and in quite a large number of cases spectrometers have both methods of automatic frequency locking available.

3.6 Magnet requirements

Most electron spectrometers employ one of two wavelengths and these correspond to the two readily available radar bands centred on 3.2 cm and 8 mm wavelengths respectively. It follows that the magnetic fields required to observe electron resonance will either be centred around 0.33 tesla (3,300 gauss) in the first case, or around 1.3 tesla (13,000 gauss) in the second case. The magnets which are employed in electron resonance spectrometers must therefore be capable of producing these fields without any difficulty, and if transition complexes are to be studied, which may have 'g' values which drop considerably below 2.0, then much higher values of magnetic field will be required. The other main feature of the magnetic field is its uniformity, or homogeneity, and any lack of homogeneity in the field over the volume of specimen itself will produce artificial broadening of the observed resonance lines. In applications in physical chemistry, where free radicals are being studied, very fine splittings and hyperfine splittings can be observed in the spectra which are only of the order of 25 milligauss or less. It then follows that the homogeneity of the magnetic field over the volume of the specimen must be better than one part in 10^6 or one part in 10^7 and hence magnet pole pieces must be very precisely engineered.

In most applications to the study of magnetic material, however, the linewidths are not so small and hence the homogeneity is not such a major consideration. This can be quite an important point if cost limitation applies to the purchase of equipment, since magnets with a reasonable homogeneity, suitable for the study of magnetic material, can cost very much less than those with the highest homogeneity required for free radical, and other studies of very fine splittings.

Although permanent magnets are often employed in nuclear magnetic resonance studies they are not normally used in electron resonance spectrometers. In nuclear resonance all spectra associated with a given nucleus can be expected to appear very close to a single magnetic field value provided constant frequency is employed. This is certainly not the case in electron resonance unless the equipment is going to be used for free radical studies, and it is normally imperative to employ an electromagnet, so that the field can be changed easily and rapidly. As well as homogeneity in space, uniformity of the magnetic field in time is also required since high sensitivity spectrometers will require long times for recording a given spectra and it is essential there are no random changes in either magnetic field, or klystron frequency, during this operation. Thus it is normally necessary to employ carefully designed stabilizing systems to keep the current which feeds the magnet coils at a constant value.

A recent development in the use of magnets for electron resonance spectrometers has been the application of superconducting magnets for this purpose. These are particularly useful if large zero field splittings exist within the compounds being studied, and they are also necessary if spectrometers are built to operate in the shorter wavelength, higher frequency, region of 4mm and 2mm wavelengths respectively. Considerable progress has been achieved in the design and operation of these magnets in recent years, and it is now possible to purchase a superconducting magnet that will give a field of 5 tesla (50 kilogauss) or so, with a homogeneity which is quite acceptable for most electron resonance purposes. The spectrometers employing such magnets are still being used mainly for specialist measurements however, and most studies on straightforward magnetic material can normally be carried through with a standard 3cm or 8mm wavelength spectrometer as normally available.

4. E.S.R. of Paramagnetic Specimens

The particular ways in which electron resonance can be employed to study ferromagnetic, anti-ferromagnetic and ferrimagnetic materials will be discussed in the next section, and although it is probable that most readers will find these particular measurements of most relevance to their own interest, it is probably best to consider the particular parameters associated with the more straightforward paramagnetic absorption first. The way in which these are modified by the internal fields of the ferromagnetic type materials can then be considered in more detail later. This section therefore summarises briefly the different parameters which are associated with electron resonance spectra from paramagnetic type material under five separate headings.

4.1 Integrated intensity of absorption line

The first parameter that can be used to characterise an E.S.R. absorption is the integrated area under the absorption curve. It can be shown that under certain specified conditions, which avoid saturation, the integrated intensity of the absorption line is directly proportional to the number of unpaired electron existing in the sample. In fact a specific expression can be derived for power absorbed in the sample which may be written

$$P_{abs} = \frac{\pi N_o}{8} \cdot \frac{\left(g\mu_B B_1 \omega\right)^2}{kT} \cdot g(\omega-\omega_o) \qquad 4.1$$

This expression contains parameters measuring the microwave magnetic field strength, B_1, and the microwave frequency, ω_1, as well as the temperature of the specimen, T. It also includes the lineshape function $g(\omega-\omega_o)$ which represents the normalised shape of the absorption curve, or in other words the integrated intensity under the line.

The normalised line shape can be abbreviated to a parameter, $\Delta\omega$, representing the half-width of the absorption line, as explained in the section 4.2.2 The minimum number of unpaired electrons that can be detected by the spectometer can then be written as

$$N_{o\,min} = \frac{4kT}{g^2\mu^2_B} \cdot \frac{\Delta\omega}{\omega_o} \cdot \frac{1}{Q_o \cdot \pi \cdot \eta} \cdot \left(\frac{F.kT.\Delta\nu}{P_o}\right)^{\frac{1}{2}} \qquad 4.2$$

specimen properties · resonance condition · cavity properties · detecting system

The determination of the number of unpaired electrons in the specimen may not be of particular interest in certain applications, but on the other hand in such fields as radiation damage studies, as well as the detection and quantitative estimation of small amounts of magnetic material, it can be invaluable. In practice the actual number of unpaired electrons present is normally not calculated directly in absolute terms from the above equation, but is obtained by comparison with a known standard which is placed in the spectrometer at the same time and under the same conditions as the specimen under investigation.

4.2 Linewidth

The second parameter associated with a single absorption line is its width. Thus a line width of given integrated area can either be sharp and narrow, or broad and shallow, and it will be evident that this second parameter reflects the spread in energy across the levels of the unpaired electron itself. In other words the measurement of the width of the absorption line will also give a measurement of the interactions between the unpaired electrons and its surroundings. In general such interactions can broadly be divided into two groups, those in which the interaction is directly with the lattice or rest of the molecule or specimen as a whole, and which is termed "spin-lattice interaction", and then secondly interactions between the spins themselves, known as "spin-spin interaction".

4.2.1 Spin lattice interaction

The theory of the actual interaction between the unpaired spins and the thermal vibrations of the lattice, or molecular array, has developed through various stages over recent years. Two types of coupling were initially suggested, the first by Waller[7] which acted via the magnetic dipole coupling between the spins themselves, and the spatial vibrations of the paramagnetic ions. This magnetic coupling is essentially very weak, however, and the second mechanism suggested by Kronig[8] acted via the internal crystalline electric field, which is altered by the thermal motion of the negatively charged electrons. The final coupling to the energy of the spins then takes place via the spin-orbit coupling, which is thus the parameter which comes into all expressions deduced from this basic theory. This mechanism envisages two processes whereby the coupling can actually take place. The spins can either exchange a whole quantum directly with a lattice vibration of the appropriate frequency and this is known as the "Direct Process". Alternatively the electron can scatter a quantum of the lattice and change its value in a "Raman type process" and, since all the quanta can take part in such a process, this is the one that is likely to predominate at the higher temperatures. Kronig in fact made an estimate for the order of magnitude of relaxation times which might be expected from these two processes and obtained expressions of the form

$$\text{Direct Process:} \quad \text{Relaxation time} = \frac{10^4.\Delta^4}{\lambda^2.B^4.T} \text{ secs} \qquad 4.3$$

$$\text{Raman Process:} \quad \text{Relaxation time} = \frac{10^4.\Delta^4}{\lambda^2.B^2.T^7} \text{ secs} \qquad 4.4$$

where Δ is the height of the next orbital state above the ground state measured in cm^{-1}, λ is the spin orbit coupling coefficient and T is the absolute temperature.

The main predictions of these two expressions are confirmed by experiment, since paramagnetic ions with small values of Δ generally have broad linewidths which vary markedly with temperature, indicating a strong spin-lattice relaxation. Moreover, when detailed quantitative measurements are made of the variation of the spin-lattice time with temperature, two clearly defined parts of the curve are generally observed one corresponding to an inverse temperature relation of the Direct Process at low temperature, while at higher temperatures a much higher rate of change with temperature occurs.

4.2.2 Spin-spin interaction

The other basic relaxation mechanism is that which exists between the unpaired electron spins themselves. Each such unpaired spin, whether it be attached to a molecule in a free radical, or to an ion of a single atom, may be regarded as a magnetic dipole, which will be precessing in the applied magnetic field. Its component in the direction of the field will have a steady value and this will produce an additional magnetic field at the site of the neighbouring unpaired electrons. The total value of the magnetic field seen by them is thus shifted slightly, and the value of this shift will vary markedly with the angle between the applied field and the line joining the two electron spins being considered. The angular dependence takes the general form associated with the magnetic field produced by a small dipole and varies as $(1-3\cos^2\theta)$. Different neighbouring unpaired electrons will thus experience different additional fields, and hence their energy levels will be shifted slightly and a general broadening of the observed absorption will occur.

As well as the simple dipole-dipole effects there will also be an additional broadening, which is produced by a rotating component of the precessing electron. If this rate of precession is the same as that of neighbouring unpaired electrons (i.e. they have the same g-value) the oscillating field which is set up will induce transitions in the neighbouring electrons and thus decrease the normal lifetime of their excited energy state. This increases the natural linewidth and hence a further broadening is produced. These two effects combine to give an expression for the mean square width of the line of the form

$$(\Delta B)^2 = \frac{3}{4} S(S + 1) (g \mu_0 \mu_B)^2 \sum \left(\frac{1-3 \cos^2\theta}{r^3} \right)^2 \qquad 4.5$$

$$\left[(\Delta H)^2 = \frac{3}{4} S(S + 1) (g \mu_B)^2 \sum \left(\frac{1-3 \cos^2\theta}{r^3} \right)^2 \right]_{\text{emu}}$$

To a first approximation this broadening and interaction is independent of both the temperature and the magnitude of the applied field and the only way in which it can be reduced is by increasing the distance between the spins (i.e. the value of "r" in the formula above). In practice this means diluting the specimen with an isomorphous diamagnetic compound. When such dilution has been carried out, so that the distance between the unpaired electrons is considerable, a dipolar broadening effect may then be found from the magnetic moments of the surrounding nuclei. This is particularly noticeable in hydrated salts where the nuclear magnetic moment of the protons surrounding a paramagnetic ion can produce a linewidth of about $6\text{x}10^{-4}$T. The only way in which such a broadening can then be reduced in the solid state is by replacing the protons with deutrons and growing the crystals out of heavy water. It should be noted here, however, that the dipole broadening can be removed completely if rapid motion of the molecules takes place as normally occurs in the liquid state. The value of the $(3 \cos^2 \theta-1)$ variation is then averaged to zero in a time short compared with the inverse of the linewidth frequency and as a result the microwave absorption only registers the average field value for the resonance absorption line. This phenomenon of motional narrowing is, of course, one of the crucial factors necessary for the observation of high resolution electron resonance spectra in solutions.

There are other similar interactions which can produce a narrowing of the absorption lines by rapidly interchanging the unpaired electrons. The most striking of these is called "Exchange Narrowing" and it occurs when electrons can be exchanged rapidly between the orbitals of different molecules. Provided this is fast enough, an averaging of the magnetic field which they experience will then take place. The detailed theory of such exchange interactions has been treated by several authors and it has been shown in general that if exchange is between similar ions or molecules then it will narrow the absorption line in the centre and broaden it in the wings, leaving the second moment unchanged. The onset of such exchange or motional narrowing can therefore be seen from a change in shape of the absorption line. If only the normal dipole-dipole interaction is present the absorption line will have a Gaussian shape, as indicated in Figure 12 while the onset of Exchange Narrowing will produce a line with a Lorentzian shape with the features as shown in Figure 13. These two line shapes can in fact be characterised by the properties summarized in the following table.

Gaussian Line	Lorentzian Line
Normalised Equation	
$g(\omega-\omega_o) = \frac{T_2}{\pi}\exp\{-(\omega-\omega_o)^2 T_2^{\,2}/\pi\}$	$g(\omega - \omega_o) = \frac{T_2/\pi}{1 + (\omega-\omega_o)^2 T_2^{\,2}}$
Width at half height	
$(\pi .\ \log_e 2.)^{\frac{1}{2}}\ \frac{1}{T_2}$	$\frac{1}{T_2}$
Width at point of maximum slope	
$\frac{1}{T_2}\sqrt{\frac{\pi}{2}}$	$\frac{1}{T_2\sqrt{3}}$

Note: In these equations T_2 is defined as the inverse of the linewidth parameter and is equal to π times the maximum value of $g(\omega-\omega_o)$.

It will be seen that both spin lattice and spin spin interactions can broaden the absorption line, but it is possible to differentiate between them in a variety of ways. Detailed measurements on the width of the absorption line can thus give information, not only on the couplings of the electrons to the lattice as a whole, but also on the interactions between the unpaired electrons on various atoms, which can be of particular significance when discussing the nature of the bond or interaction between them.

4.3 Resonance condition and g-value

The third parameter which can be associated with this single absorption line, is the actual resonance condition for which it occurs. Since the applied microwave frequency is normally held constant in all cases studied, the only variable on the experimental side is the value of the applied magnetic field , B Reference to the basic resonance condition of equation 1.1 will show that since all the other parameters,such as μ_B and h are constants, the only other variable in the expression is the value of g. If the unpaired electron were entirely free, and did not interact at all with the orbital momentum of any atom with which it was associated, the value of g would, in fact, be the free spin value of 2.0023 (the additional 0.0023 coming from the interaction of the electron with its own radiation field). In a large number of free radical studies, and others associated with highly delocalised electrons, the g-value is found to be exceedingly close to the free spin value, indicating that the electron is moving in a highly delocalised orbital. However, in transition group complex studies, and others in which there is fairly strong bonding between the one electron and a single atom, this g value can shift very considerably from the free spin value, and this shift reflects the strength and nature of the chemical bonding in which the electron is taking part. It therefore follows that a determination of this g-value, and in particular a determination of its angular variation, can often give very precise information on the nature of the chemical bond around the atom in question, and also on the details of the higher energy levels of the particular atomic, or molecular, configuration.

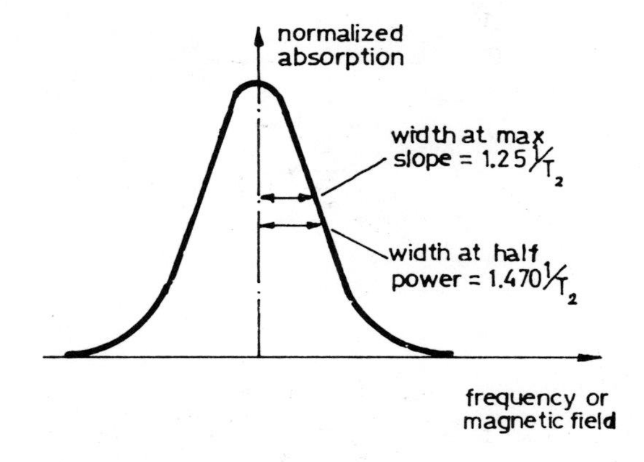

Figure 12 GAUSSIAN LINE SHAPE

Comparison with the Lorentzian shape of Figure 13 shows that this is characterised by a larger half-width but with less absorption in the wings.

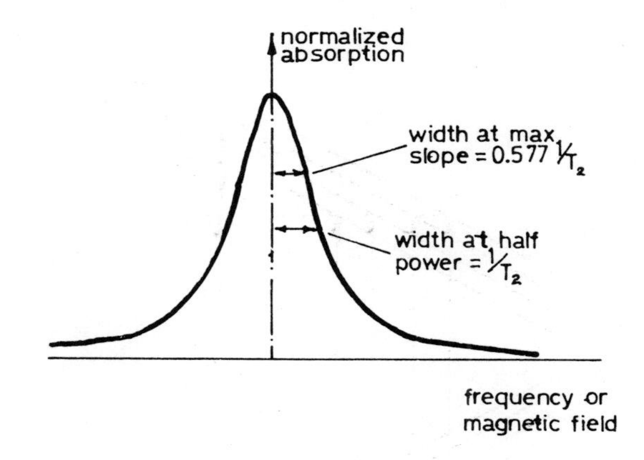

Figure 13 LORENTZIAN LINE SHAPE

This shape is observed whenever stray exchange or motional narrowing is present.

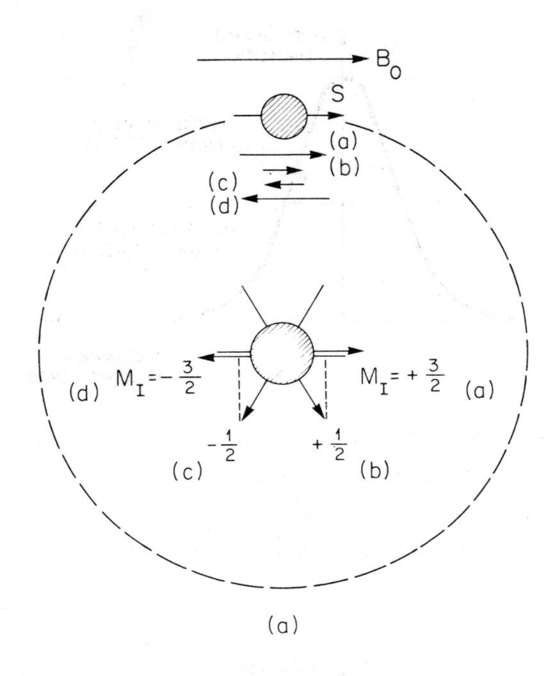
B0
S
(a)
(b)
(c)
(d)
(d) $M_I = -\frac{3}{2}$
$M_I = +\frac{3}{2}$ (a)
$-\frac{1}{2}$
(c)
$+\frac{1}{2}$
(b)

(a)

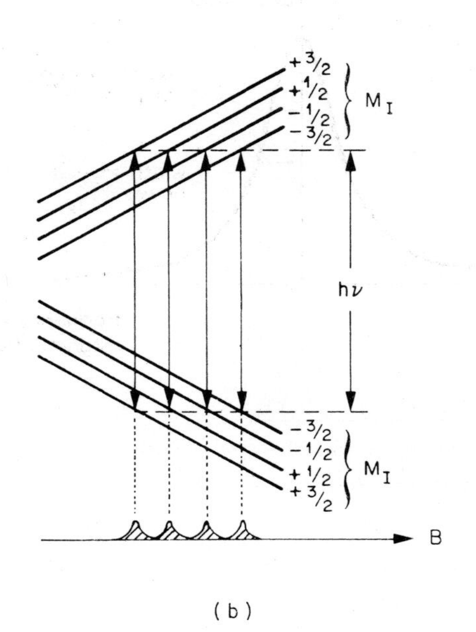
+3/2
+1/2
−1/2
−3/2
M_I
hν
−3/2
−1/2
+1/2
+3/2
M_I
B

(b)

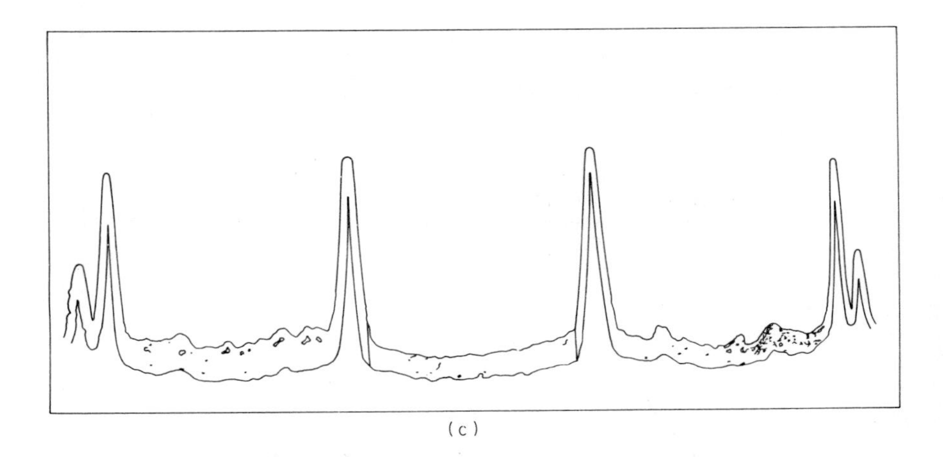

Figure 14 HYPERFINE INTERACTION FROM COPPER NUCLEI

a) Four possible orientations of the copper nucleus and resultant incremental magnetic fields.

b) Resultant splitting or electronic energy levels into four components.

c) Observed hyperfine splittings. Two overlapping sets of four lines are produced by the two copper isotopes Cu^{63} and Cu^{65}.

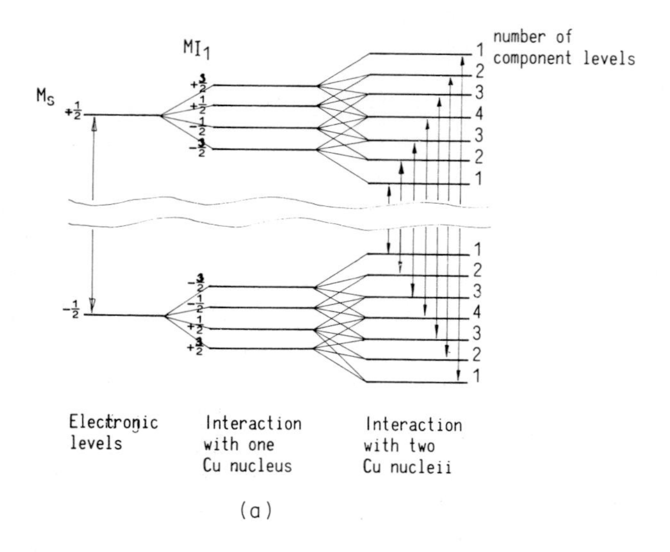

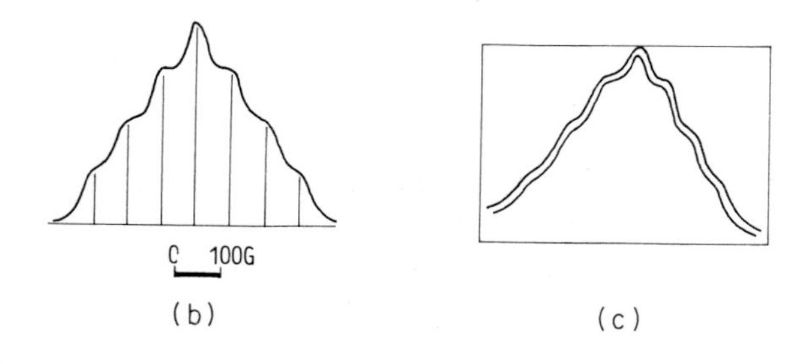

Figure 15 HYPERFINE INTERACTION FROM TWO EXCHANGE-COUPLED COPPER IONS

a) Energy level splitting expected for an electron equally coupled to two copper nuclei.

b) Predicted overall hyperfine pattern.

c) Observed hyperfine structure.

4.4 Hyperfine splitting

As well as these three parameters associated with a single absorption line there are two additional effects which will cause the single line to be split, and these splittings can in turn give very interesting additional information. These splittings arise from internal magnetic fields which exist within the specimen and act on the unpaired electron in addition to the applied external magnetic field. Hence, if any internal field has the possibility of more than one steady value it is possible for the actual resonance position to be shifted from one value to another as the internal field adds to, or subtracts from, the applied field. One of the most obvious sources for such an internal magnetic field is the nucleus of the atom itself. A large number of nuclei do in fact possess nuclear spins and magnetic moments, and although these magnetic moments are very small, being about 2,000 times smaller than that of the electron, the field which they can produce at the site of their electrons is nevertheless quite large, owing to the small distances involved. In fact a rough classical calculation will indicate that the field produced by one nuclear magneton at the position of a typical electron orbiting that atom is of the order of 0.01 tesla (100 gauss), and a splitting of the spectra of this magnitude can be observed extremely easily.

It will probably help if a definite example is taken of such an electron-nuclear interaction, and the particular case of the unpaired electron in a 3d orbit of a copper atom will serve to illustrate this point. This situation is represented in Figure 14. The copper nucleus itself has a nuclear spin of $\frac{3}{2}$ and hence it can take up four possible orientations in the applied external field B_0. These four possible orientations arise from the four possible resolved magnetic moments in the direction of the field of $+\frac{3}{2}$; $+\frac{1}{2}$; $-\frac{1}{2}$; and $\frac{3}{2}$ as indicated in Figure 14 (a). The unpaired electron, as it moves around the copper nucleus, will now not only see the applied magnetic field B_0, but also small incremental fields arising from the copper nucleus. The magnetic moment of the copper nucleus may, in fact, be in any of the four possible orientations we have considered, and therefore the electron may have any of the four additional increments as represented by the small arrows at the top of Figure 14a. Their effect on the energy level diagram is shown in Figure 14b, where the two original diverging levels are now replaced by two sets of four levels, these four levels corresponding to the four possible orientations of the copper nucleus. The constant energy quantum, represented by $h\nu$, will now cause transitions between these levels at four different values of the magnetic field, as indicated by the four shaded lines below. We therefore expect the electron resonance spectrum from copper to consist not of a single line, but of four hyperfine components of equal intensity. The spectrum actually obtained is shown in Figure 14c, and the four component lines can be clearly seen. These are actually all doubled, although in the two central lines the separate doublets are not resolved. This doubling arises from the fact that copper contains two abundant isotopes Cu^{63} and Cu^{65}; both of these have a nuclear spin of $\frac{3}{2}$ but they have slightly different magnetic moments and hence produce slightly different splittings. The sensitivity of this technique, in not only producing well-resolved hyperfine patterns, but also enabling isotope analysis to be carried out even when the isotopes have the same nuclear spin, is demonstrated very clearly in this figure. It follows, in general, that for a nucleus with nuclear spin I the hyperfine pattern will consist of $(2I + 1)$ lines of equal intensity and equal spacing. It is evident that this can be used as a very straightforward method of identification, the known hyperfine pattern being used as a "characteristic fingerprint" of the unknown atom.

4.4.1 Information on exchange coupling from hyperfine splittings

The above section has discussed the hyperfine interaction which arises when the unpaired electron is interacting with only one nuclear magnetic moment. It is possible, however, for the unpaired electron to be coupled to more than one nuclear moment and more complex hyperfine structures are then observed which can, in turn, give detailed information on the wavefunction of the unpaired electron, and on any interactions within the solid which cause the electron to be shared between different atoms. There are many different examples of the complex hyperfine splittings that can be obtained from such multi-nuclear interactions but probably the most relevant to the present article are those where direct exchange interaction is taking place between the atom under consideration and others which are relatively close in the crystal lattice.

As an introduction to the type of spectrum that is obtained under such conditions, the case of copper can be followed through to a salt where the copper atoms are strongly coupled in pairs, and the best known example of this is copper acetate.[9] In this crystal the structure is such that two copper ions are situated close together and the two unpaired spins on these two ions are exchange-coupled to form a singlet ground state with $S = 0$ and a triplet excited state $S = 1$ some 300 cm^{-1} above it. The population of the higher triplet state therefore depends quite crucially on the temperature of observation, and at room temperature there is sufficient excitation into it to give a high concentration of triplet states which produce a broad line in which no hyperfine structure is resolved. On the other hand, if the crystal is cooled to $77^{o}K$, the concentration in the triplet state is considerably reduced and, as a result, the broadening is diminished and the hyperfine pattern is observable. On the other hand, if the salt is cooled below $20^{o}K$, all the ions are in the ground state, with paired spins, and no electron resonance absorption is observed at all.

The hyperfine pattern which is observed at $77^{o}K$, however, is not the equally intense four-line pattern discussed in the last section, but one which has seven components of relative intensities 1:2:3:4:3:2:1 which is exactly the pattern which would be expected from an electron coupled equally to two copper atoms with a spin $I = \frac{3}{2}$, as illustrated in Figure 15. The existence of this particular splitting is therefore unequivocal proof that the strong exchange interaction exists between the two copper ions, and the variation in intensity of the electron resonance signal with temperature can itself be used to determine the strength of the exchange coupling quite directly.

A large number of other salts which have anti-ferromagnetic coupling have also been studied in this way, and K_2IrCl_6 is a good example of such a compound, that has been extensively studied by electron resonance.[10,11] Thus it is possible to grow magnetically dilute single crystals containing about 5% of the K_2IrCl_6 dispersed in the diamagnetic host lattice of potassium chloro-platinate, i.e.

K_2PtCl_6. This is still a fairly high concentration of the paramagnetic salt, and there are still a large number of ion pairs which occupy adjacent lattice sites. Each of these two iridium ions which are in the d^5 state, have a $S = \frac{3}{2}$ and hence the combined pair gives a singlet and triplet state with a small zero field splitting. The magnitude of this splitting, which itself directly determines the exchange coupling constant, can again be deduced from the intensity variation of the e.s.r. spectrum with temperature. It is also found that the hyperfine structure which is observed has contributions from the surrounding chlorine atoms, as well as from the iridium nuclei themselves. This type of information which can be deduced directly from the electron resonance spectrum is of very great value in understanding the actual mechanism of the anti-ferromagnetic interact-tion in such salts as these. Thus the exchange coupling is not between the two iridium ions directly, via an overlap of their d-orbitals, but is instead transmitted through the intervening chlorine ligands and wavefunctions , and is an example of a "super exchange" or "indirect exchange".

This example is only one of very many which have now been studied by electron resonance in this way, but it will serve to indicate the detailed information that can be deduced on the mechanism of ferromagnetic and anti-ferromagnetic exchange forces. Although sometimes the measurements on diamagnetically diluted material are often the ones most closely investigaged , the results can often be extrapolated very directly back to the concentrated magnetic material itself, to give information on the exchange interactions present in this.

4.5 Electronic splitting

The fifth and final parameter to be associated with electron resonance spectra is the electronic splitting, which can occur if more than one unpaired electron is associated with a single atom. The particular case of an atom with two unpaired electrons associated with it may be taken as an example. In this case the two electrons will couple to give a resultant total spin quantum number of S = 1, as has already been discussed in the case of the exchange coupling in the previous section. This overall spin quantum number can now take up three different possible orientations in the applied magnetic field, with components M_s = +1, 0 or -1. If the atom is in free space and has no electric or magnetic fields acting on it, these three orientations will all have the same energy , but an applied magnetic field will separate them as shown in Figure 16a. It will be seen that incoming microwave quanta will always produce two transitions at exactly the same resonance field value in such a case, and thus only one single absorption line would be observed. If, however, the atom is located within a crystal, it will have a strong internal electric field acting on it. This internal electric field will produce a Stark effect and separate the M_s = 0 from the M_s = ± 1 levels, even in the absence of any applied magnetic field. The situation , therefore, changes to that shown in Figure 16a. It will be seen that incoming microwave quanta will always produce two transitions at exactly the same resonance field value in such a case, and thus only one single absorption line would be observed. If, however, the atom is located within a crystal, it will have a strong internal electric field acting on it. This internal electric field will produce a Stark effect and separate the M_s = 0 from the M_s = ± 1 levels, even in the absence of any applied magnetic field. The situation, therefore, changes to that shown in Figure 16b. When the magnetic field is applied in this case the levels will diverge as shown, and it is now evident that incoming microwave quanta will produce resonance absorption at two different values of the applied magnetic field strength. Two absorption lines are thus obtained, and the splitting between these , in fact, reflects the splitting between the M_s = 0 and M_s = ± 1 levels, in zero magnetic field. This type of splitting is termed an "electronic splitting", since it arises from the different orientations of the total electronic magnetic moment. In general it is orders of magnitude greater than the hyperfine splitting discussed previously and thus can be very clearly distinguished from it.

A determination of the magnitude of the electronic splitting, from the electron resonance, can often give precise information on the fields existing within the crystal, a factor which can be of crucial importance in understanding the mechanism of exchange, or other solid state, interactions.

5. Ferromagnetic Resonance, Anti-Ferromagnetic Resonance and Ferrimagnetic Resonance.

5.1 The investigation of exchange interactions

It has already been seen that the crucial difference between electron paramagnetic resonance and the various types of ferromagnetic resonance that can be observed is the existence in the latter of strong exchange interactions , which couple the unpaired spins together in the solid and produce strong additional internal magnetic fields. The investigation and better understanding of these exchange interactions themselves is therefore a crucial part in the development of any basic theory of ferromagnetism, and it is in this connection that electron resonance has probably made its greatest contribution to the study of magnetic materials. The way in which such information can be obtained has already been discussed briefly when considering the complex hyperfine patterns that are obtained from salts in which exchange coupling is present, as discussed in section 4.4.1.

Another example which is often taken as a basic model for anti-ferromagnetic interactions is the case of MnO. Most of the proposed "super exchange" mechanisms in this crystal involve spin transfer between the atoms via the intervening oxygen atom[12], and can be represented diagrammatically as in Figure 17. The typical orbits on both the metal atom, M, and the intervening ligand atom, X, (in this case an oxygen atom) are shown in this diagram. The mechanism of the interaction must therefore be such that it can mix some of the electron orbit of the $d_{3z^2-r^2}$ on the manganese ion on the left, with a similar orbit on the manganese ion on the right. These particular magnetic orbits are in fact anti-bonding orbits which contain small admixtures from the $2p_z$ orbit of the intervening oxygen ion. In the first detailed treatment of this Anderson considered the appropriate anti-bonding levels which could be admixed to give appropriate molecular orbitals, and was able to deduce a theoretical expression for the energy by which the singlet ground state was depressed, and which can thus be attributed to the anti-ferromagnetic interaction between the two spins. Comparison of his theoretically predicted value

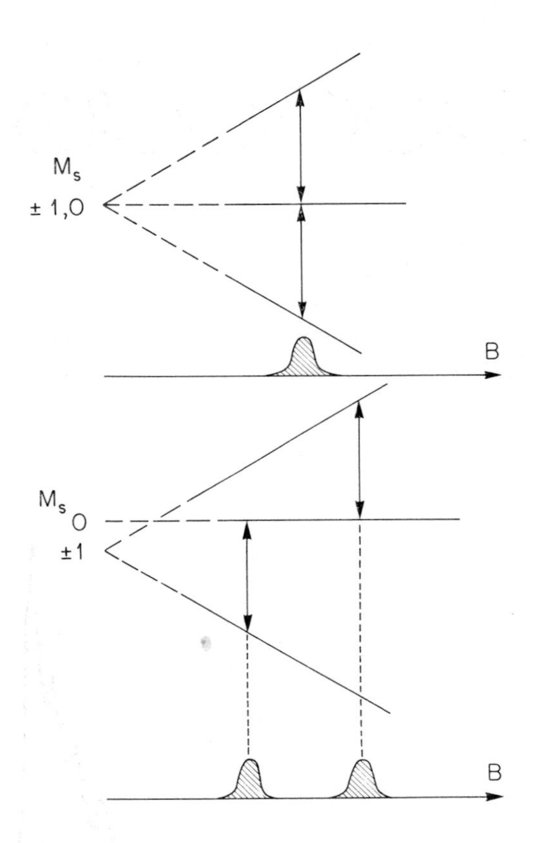

Figure 16 ELECTRONIC SPLITTING

a) Production of three energy levels, but only one absorption from two overlapping transitions, when no zero-field energy splitting present.

b) Zero-field splitting produced by internal crystalline field produces resultant splitting in observed e.s.r. absorptions.

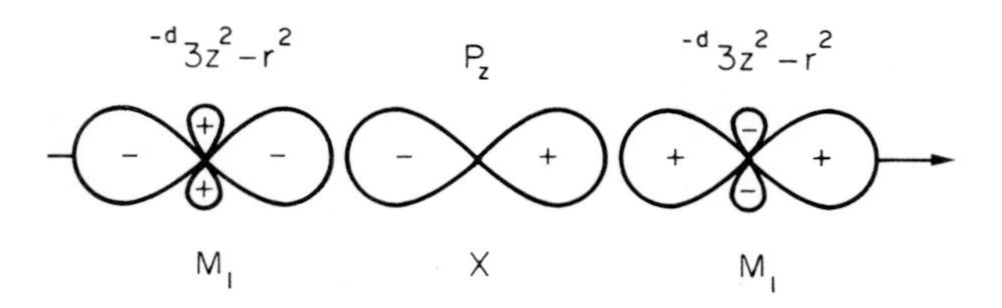

Figure 17 MODEL FOR SUPER-EXCHANGE IN MnO

The $d_{3z^2-r^2}$ orbitals have been drawn on the Mn^{2+} ions and the P_z orbital on the intervening oxygen atom to show how these form a 180° super-exchange path between the two metal ions.

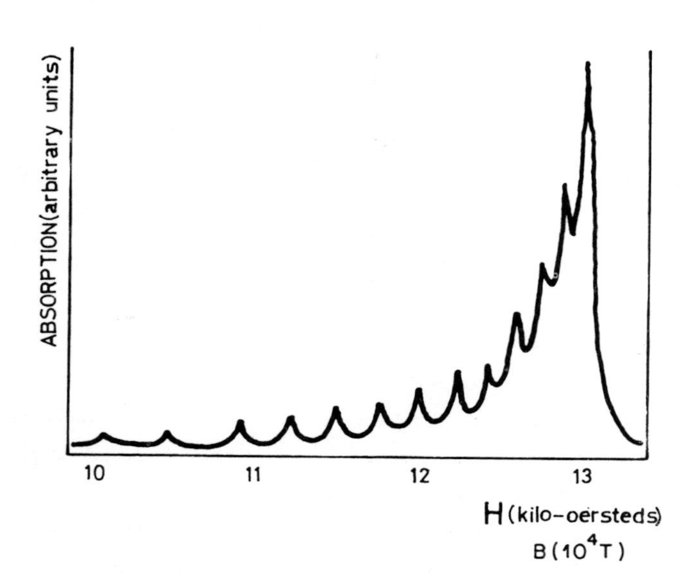

Figure 18 SPIN-WAVES IN A PERMALLOY FILM

The external d.c. field was applied normal to the plane of the film and hence the different resonances correspond to different values of k in equation 5.6.

with those observed experimentally for the exchange coupling constant give an order of magnitude agreement which was considered quite satisfactory by Anderson in his original treatment. However a more important result from this theory is the fact that they lead to some very general predictions for such "super exchange" interactions, and these can be summarised briefly as:

(i) If two magnetic ions can transfer unpaired spins into the same ligand orbit, the spins will try to couple anti-parallel by Pauli's principle, so that there is an antiferromagnetic contribution to the exchange interaction. The magnitude of this is proportional to the probability of finding the spins simultaneously in the ligand orbit, that is to the product of the relevant spin transfer coefficients. The second general point that can be made refers to the ferromagnetic contributions to the exchange, which are usually smaller than the antiferromagnetic ones discussed above, but which are governed by a similar rule which can be stated as follows.

(ii) If two magnetic ions can transfer unpaired spins into different orbits on the same ligand, the spins will try to couple parallel by Hund's rule, and there is a ferromagnetic contribution to the exchange interaction. The magnitude of this is again proportional to the probability of finding these spins simultaneously on the ligand, that is to the product of their relevant spin transfer coefficients.

These general statements can now be applied to various specific cases to predict the proportions of anti-ferromagnetic and ferromagnetic coupling that will be found. Moreover the ability to check such predictions quite precisely by the experimentally observed electron resonance spectra enables a direct check to be made on the theory and predicted magnitudes of the various types of exchange which exist in magnetic materials.

5.2 Ferromagnetic resonance

As has already been mentioned in the introductory section, the first observations of ferromagnetic resonance itself were made by Griffiths[2] in 1946 and a few months later Kittel[3] produced a theoretical treatment which accounted very well for the general features of this resonance including the shift of the resonance field value from those expected for free electron spin. Thus to a first approximation it is possible to derive an expression for the susceptibility component in the direction of the microwave magnetic field χ_x

$$\chi_x/\chi_o = \left(1 - \frac{\omega}{\gamma(B_{m_z} B_{o_z})^{\frac{1}{2}}}\right)^{-1} \qquad 5.1$$

and it will be seen that this predicts that the permeability would rise to an infinite value at the resonance condition given by

$$\omega = \gamma(B_{m_z} B_{o_z})^{\frac{1}{2}} \qquad 5.2$$

which is the same thing as saying that the resonant field value, which is normally the externally applied field B_o is now replaced by $(B_o B_m)^{\frac{1}{2}}$.

5.2.1 g-values

Early measurements showed that this simple theory was in fact a good first approximation, as can be seen from the table of results below where the experimentally observed resonance frequencies are compared with those predicted for the two cases of the simple external field B_o and the product of the external field and the internal flux.

Table 5.1 Experimental and Calculated Resonant Frequencies $\left(\frac{10^{10}}{2\pi}\right)$ Hz

Ferromagnetic	D.C. Field B_o	Calculated Lamor (i) for B_o	Frequencies (ii) for $(B_o B_m)^{\frac{1}{2}}$	Experimental Frequency
Fe	2,800	5.0	14.5	15.4
	500	0.9	5.8	5.9
Co	510	0.9	5.3	5.9
Ni	5,000	8.8	13.5	15.4
	3,800	6.7	10.9	13.2
	1,030	1.8	4.9	5.9

This simple theory, however, takes no account of the various relaxation mechanisms which are present in a magnetic material, and these can be in fact of very considerable magnitude. The effect of these will be to add a damping term to the original classical equations describing the motion of the spins, and the expression for the susceptibility then also includes a term in which these relaxation forces are included. Another effect which must also be added to the total consideration is the effect of the demagnetising field due to the actual shape of the specimen itself, and also any effects due to the anisotropy of the magnetic properties of the bulk of the material. The condition for resonance in the general case then becomes extremely complicated, in which relaxation effects, anisotropy effects and demagnetisation effects must all be included. As a result of this, experimenters have tended to concentrate measurements on one of three specific types of arrangement, so that the results can be fairly easily interpreted. These three cases can be briefly summarised as follows:

Case 1 The applied external magnetic field B_0 and the microwave magnetic field are both applied tangentially to a thin film of the ferromagnetic (as was the case in the original experiment of Griffiths). Two of the demagnetising coefficients are then zero, and the condition for resonance reduces to the simple equation

$$h\nu = g.\mu_B(B_oB_m)^{\frac{1}{2}} \qquad 5.3$$

which was that originally derived by Kittel.

Case 2 This is basically the same arrangement as for Case 1 above, but with the applied magnetic field normal to the sheet of ferromagnetic material. The two demagnetising factors in the plane of the specimen now become zero, but the one along the direction of the applied field takes on a finite value, and as a result the condition for resonance becomes

$$h\nu = g\mu_B \, (B_o - \mu_oM) \qquad 5.4$$

This is again a fairly straightforward equation to apply, but it should be noted that the thickness of the ferromagnetic sheet must be less than the skin depth if this simple relation is to hold.

Case 3 In this case the sample is formed by a collection of very small spheres of magnetic material which are placed in the normal configuration, with the external magnetic field perpendicular to the applied microwave field. In this case all the demagnetising factors become equal, and the resonance condition reduces to the straightforward equation for normal paramagnetic resonance, i.e.

$$h\nu = g\mu_B B_o \tag{5.5}$$

In most experimental work it has been found easiest to employ the arrangement of case 1 above, but the validity of the equations of the other two cases have been checked experimentally as in the early work of Kittel, Yager and Merritt,[13] who observed the shift in the absorption curve of an annealed supermalloy specimen, when the applied field was changed from the parallel to the perpendicular direction. The ideal experimental specimens are in fact those corresponding to case 3, since no demagnetising factors need to be calculated, and a very direct measurement of the spectroscopic splitting factor g is obtained. These spheres must have diameters less than the skin depth, however, if the equation is to be applied strictly, and the normal way of obtaining such specimens is by the production of colloidal suspensions in paraffin wax, containing the spherical particles, which must be under about 100A in diameter.

Following the early measurements which served to verify the basic ideas of the initial theory, the experimental work on ferromagnetic resonance was then concentrated on studies of magnetic anisotropy, using for example single crystals of iron.[14] These anisotropic effects can be very considerable , and can, for example, even produce additional resonance lines owing to a competition between the internal magnetic fields. Thus, if the external field is applied along a direction of difficult magnetisation, there will be opposing forces between the applied field and the field set up along the easy direction of magnetisation within the specimen itself. If low microwave frequencies are being employed (e.g. 9,000 MHz), so that the normal resonant field value is not very high, the direction of magnetisation will not follow that of the applied field initially and hence the field required for resonance will be reached while its direction is still determined by the anisotropy axis of the crystal rather than the direction of the applied field itself. As a result, an absorption will be obtained at a relatively low field value. As the magnitude of the external field is further increased, however, the direction of magnetisation within the specimen is in effect pulled round to the direction of the applied field, and absorption will again occur at a value predicted by the resonance equation suitable for that particular configuration. It is probably fair to say that the agreement between the theoretical treatment of these anisotropy forces and the measured resonant field values, is one of the most satisfactory areas of ferromagnetic resonance theory, and the anisotropy constants of particular crystals can often be predicted in detail from the resonance measurements themselves.

5.2.2 Linewidths

In contrast to the good agreement between the predicted g-values, or resonance field value, for the ferromagnetic resonance absorption lines, the explanation of the observed linewidths of such resonances has been much more difficult to account for completely. Thus at first sight it might be thought that the ferromagnetic resonance absorption lines would be expected to have very large widths as the electron spins are very close together, and hence the ordinary dipole broadening between them should be quite large. If this were the only internal interaction present, such an explanation should be correct, but the presence of a large exchange interaction which must, of course, be much larger than for ferromagnetics than for any paramagnetic case in which it has been observed, would suggest that very much narrower absorption lines would be produced. In practice the linewidths observed are often many hundreds of gauss width, and hence some additional mechanisms have to be postulated which will overcome the strong exchange narrowing that would be initially expected. In the early theoretical work on such resonances various mechanisms were suggested, such as strong spin lattice interaction, the effect of anisotropic exchange forces or the damping effect produced by the skin depth effect itself.

More recently, however, it has come to be realised that one of the main additional broadening mechanisms in normal ferromagnetic resonance is probably the interaction with the conduction electrons of the metal. The importance of this becomes even more clear when it was realised that the original treatments, which assumed that the magnetic electrons responsible for the ferromagnetism of a particular ion, were in fact only a very simple approximation. Thus in the iron group of elements, it is clear that the 3d shell, which is near the outside of the ions, will have wavefunctions which overlap considerably to form a narrow band of states, so that the magnetic electrons can move from one to the other and a localised spin picture is thus no longer really applicable. Hence one would expect that there would be a very strong interaction between the magnetic electrons and the other conduction electrons themselves, and this form of relaxation is likely to be the main factor dominating the width of the observed resonance lines.

Confirmation of this fact comes from measurements on the resonance of some ferrimagnetics which are discussed in a later paragraph, where it will be seen that very narrow lines can sometimes be obtained,[15] and of course in this case there would be no additional broadening mechanism from interaction with the conduction electrons.

5.3 Spin wave resonance

Following the early work on ferromagnetic resonance itself, it was realised that the microwave resonance technique would also be the most direct method of observing spin waves in ferromagnetic material. These spin wave resonance experiments were first suggested by Kittel,[16] and first observed experimentally by Seavey and Tannenwald[17] in 1958, and the basic principle of the experiments is in fact identical to that of paramagnetic resonance in that the sample is placed in a microwave cavity with the microwave magnetic field at right angles to the externally applied d.c. field.

If, in fact, all the spins in the sample are precessing in phase with one another, then the situation is exactly as has already been envisaged for straightforward ferromagnetic resonance, and all the previous description of ferromagnetic resonance will apply to the specimen. It was realised, however, when the idea of spin waves was first conceived, that it is possible for the spins to be precessing at different phases within the specimen, and this will then give rise to different resonance conditions. In its simplest consideration, the idea of a spin wave can be visualised as the mis-orientation of one electron spin compared with the spins of all the other atoms in the sample, the latter being aligned parallel to the applied field. This mis-orientation, or deviation, will not remain static and associated with one atom, because of the strong exchange interactions between all the electrons, and hence the spin deviation can be considered as moving through the crystal lattice, and the collective excitation of this deviation moving to and fro through the lattice can be considered as a spin wave. In a formal treatment of spin wave theory, such a spin deviation may be considered as a quasi-particle and by analogy with phonons and photons, it is often referred to as a "magnon". There is no room in this particular paper for any detailed treatment of the theory of spin waves, but reference may be made to various review papers which have summarised the theoretical and experimental studies.[18] The particular relevance of spin waves to this paper, however, is that electron resonance is probably the most direct method of observing them and hence the technique does form a very useful check on the background theory that has been developed.

It has already been seen that if all the spins are precessing in phase there will be no spin deviation and the conditions of normal ferromagnetic resonance occur. If there is a phase difference between them, however, then spin waves will exist within the sample but, even so, in a bulk sample of ferromagnetic these spin waves will extend over many wavelengths, so that at any instance the transverse magnetisation across the specimen will tend to average to zero, and there will be no effective interaction with the microwave field. However, in a small specimen the number of such wavelengths will actually be limited by the sample dimensions, and at the surface of the specimen in particular spins will find themselves in a different magnetic environment from those in the body of the specimen, since there are no neighbours on one side. The surface spins are therefore said to be "pinned" and they produce effective boundary conditions, which limit the possible modes of precession, and resulting spin waves, to a standing wave pattern with nodes at the surface. Mathematical treatment of the spin waves with these boundary conditions then leads as a resonance condition for the simple case

$$h\nu = g\,\mu_B(B_o - \mu_o M) + Dk^2 \qquad 5.6$$

where D is known as the spin wave constant , or spin wave dispersion coefficient, and $k = p\pi/L$, p being an integer and L the width of the specimen.

If the applied d.c. magnetic field is not perpendicular to the specimen, this dispersion relation will be modified and, for the other form of simple geometry, in which both the applied magnetic field and the microwave field are in a plane of the ferromagnetic film, the resonance condition then becomes

$$h\nu = g\,\mu_B\left[\left(B_o + \mu_o M + \frac{Dk^2}{g\mu_B}\right)\left(B_o + \frac{Dk^2}{g\mu_B}\right)\right]^{\frac{1}{2}} \qquad 5.7$$

It can be seen however that under either of these conditions, measurement of the resonance frequencies will enable a direct measurement to be made of this fundamental constant D, which characterises the spin waves in the material. An example of an experimentally observed spin wave resonance is shown in Figure 18, which is for a thin permalloy film with the external field applied perpendicular to the plane of the film. The value of the spin wave constant, D, can then be determined with considerable accuracy from the spacing between the successive modes which are observed, and the fall away in the intensity of these can be correlated to the relaxation damping produced by the presence of the conduction electrons. When this is taken into account the full expression for the resonance condition becomes

$$h\nu = g\mu_B(B_o - \mu_o M) + Dk^2 + \frac{4\pi g\,\mu_o\mu_B M}{1 + \frac{1}{2}i\delta^2 k^2} \qquad 5.8$$

where δ is the skin depth.

The spin wave resonance measurements have been used not only to determine the absolute values of the constant, D, but also its variation in temperature in various specimens, and this variation itself can often give useful information on the validity of various models being used to explain the spin wave interaction.

5.4 Anti-ferromagnetic resonance

Although the same basic principles will apply to anti-ferromagnetic and ferrimagnetic resonance as has been considered for ferromagnetic resonance itself, it is clear that the existence of two opposing sub-lattices within the crystal structure will very considerably modify the effective fields which the individual electrons and atoms experience, and hence also modify the resonance condition itself. The detailed theory of anti-ferromagnetic resonance can be followed through in terms of two separate equations of motion for the two sub-lattices, which will also have their own anisotropy constants. There is no space in this particular paper to develop a detailed theory of antiferromagnetic resonance, but the general ideas have been developed by such workers as Yosida,[19] who has also shown that there is extremely good agreement between experimental measurements on such salts as $CuCl_2 2H_2O$, which has a fairly small exchange interaction, and the results that are predicted theoretically.

As explained earlier, it is possible for the internal fields to be of such a magnitude that resonance can be observed in the microwave region even in the absence of any externally applied field. Such measurements then enable a very direct measurement to be made of the exchange coupling constants, and extension of e.s.r. spectrometers into the mm wavelength region has considerably helped in this respect.

It is also possible to observe some such zero-field resonances in the far infrared region, and a typical summary of such measurements ranging from normal e.s.r. frequencies, through the mm wave region to the far infra-red, is given in Table 5.2.

Table 5.2 Resonance Conditions for Zero field Absorption in Anti-ferromagnetics

Compound	Frequency	Wavelength	Néel Temperature
$CuCl_2.2H_2O$	17.5GHz	1.7 cm	$4.3^{o}K$
Cr_2O_3	158 GHz	1.9 mm	307
$MnTiO_3$	150 GHz	2 mm	65
MnF_2	$8.7\ cm^{-1}$	1.15 mm	67
NiF_2	$31.1\ cm^{-1}$	0.32 mm	73
MnO	$27.6\ cm^{-1}$	0.36 mm	120
NiO	$36.5\ cm^{-1}$	0.27 mm	523

5.5 Ferrimagnetic resonance

The detailed theory of ferrimagnetic resonance can also be developed by considering separate equations of motion for the two sub-lattice magnetisations, although there will of course now be a resultant internal molecular field, and if this is sufficiently large, compared with the applied magnetic field, the resonance conditions will be very similar to those observed from ferromagnetic specimens. Thus, if the intensity of magnetisation associated with one of the sub-lattices is very much greater than that associated with the other, this will dominate the internal effective field. It is then possible for the resonance frequency , even in zero applied magnetic field, to fall within the microwave region when the resonance absorption can be observed in zero external magnetic field. It is in fact possible to derive a general expression for the resonance frequency near the compensation point of the form,

$$h\nu = g.\mu_B\left[B_o \pm \{B_A(2B_E + B_A)\}^{\frac{1}{2}}\right] \qquad 5.9$$

where H_0 is the applied magnetic field, B_E is the internal exchange field and B_A is the anisotropy field, and this has been checked experimentally by several workers. The possibility of observing such ferrimagnetic resonances in the microwave region, but in zero magnetic field, was first pointed out by Wangsness[20] and observed by McGuire[21] and by Welch et al.[22] Further work on such specimens as single crystals of gadolinium iron garnet[23] have shown that there can be very good agreement between theory and experiment in such studies of ferrimagnetic material.

It may be noted, in general, that the same kinds of experimental conditions are required for the observation of ferrimagnetic resonance, as for the ferromagnetic case, and most experiments have in fact been carried out with spherical samples, so that shape ansiotropies need not be considered. The g-value that will be observed for the resonance itself will be an average of the g-values for the individual ions in the two sub-lattices, provided there is a parallel/antiparallel arrangement between them, and the effective g-value will therefore be given by the expression

$$g = \frac{\Sigma\, g_i S_i}{\Sigma\, S_i}$$

where the summation is taken over all the ions in each sub-lattice Typical g-values that have been observed for various ferrimagnetics are listed in Table 5.3 and it may be noted that for the first two specimens listed, all the cations are in the S state and therefore a g-value close to free spin is to be expected. On the other hand, the Fe_3O_4 and $NiFe_2O_4$ compounds, which have an inverse spinel structure, will be expected to have an effective g-value which is an average between that of Fe^{2+} and Ni^{2+} ions, as is in fact observed. It follows that this procedure can also be used in reverse, and when g-values are accurately known for the individual ions which comprise the ferrimagnetic specimen, then the ionic distribution over the different lattice sites can be deduced from the measured g-values.

Table 5.3 Observed g-values in Ferrimagnetic Specimens

Specimen	Observed g-value	Reference
$MnFe_2O_4$	2.004	24
Y.I.G.	2.004	25
Fe_3O_4	2.12	26
$NiFe_2O_4$	2.19	27

This process of deriving information about the structure or nature of the magnetic material from the resonance measurements is also often used when anisotropy constants are derived, since in a large number of cases the resonance measurements give the best value of these for ferrimagnetic materials.

6. General Conclusions

Although it has not been possible in this paper to consider the detailed expressions for ferromagnetic resonance with the various spin wave modes which can be associated with it, nor the full treatment of anti-ferromagnetic resonance or ferrimagnetic resonance, it is hoped that sufficient has been given to indicate the kind of information that can be deduced from such resonance measurements, and how these can be correlated both with the basic properties of materials themselves and other measurements of bulk properties. Although it is probably fair to say that electron resonance has found its major application in the field of paramagnetic studies, where very fine splittings can be observed on the spectrum and related directly back to nuclear interactions, there is nevertheless an increasing body of knowledge which can be deduced from the resonance measurements on the ferromagnetic types of material. Although precise information on the energy levels of the individual atoms is not so accessible in these cases, because of the strong internal effective magnetic fields, there is however the compensating advantage that some information on the bulk properties of the material can be obtained from these resonance measurements, which are not normally available by such techniques. The observation and analysis of the spin waves is one example of this, and other studies on magnetostatic modes can also give very useful information on the properties of the particular crystals being studied. Further details of such work, together with more complete discussion on relaxation processes, which can be correlated with linewidths and line shapes are to be found in various more advanced reviews on different aspects of these resonance studies, which are listed in the general bibliography at the end of the references.

REFERENCES

1. A. Einstein, Phys. Zeit. 18,121 (1917)
2. J.H.E. Griffiths, Nature 158, 670 (1946)
3. C. Kittel, Phys. Rev. 71, 270 (1947)
4. L.R. Maxwell, and T.R. McGuire, Rev. Mod. Phys. 25,279 (1953)
5. D.J.E. Ingram, "Spectroscopy at Radio and Microwave Frequencies, Butterworths Ltd., Ch. 4 & 11 (1967)
6. C.P. Poole, "Electron Spin Resonance", Interscience (1967)
7. I. Waller, Zeit. Phys. 79, 370 (1932)
8. R. de L. Kronig, Physika 6, 33 (1939)
9. B. Bleaney, and K.D. Bowers, Proc. Roy. Soc. A 214 451 (1952)
10. J. Owen, and K.W.H. Stevens, Nature 171 836 (1953)
11. J.H.E. Griffiths, and J. Owen, Proc. Roy. Soc. A 226, 96 (1954)
12. P.W. Anderson, Phys. Rev. 115 2 (1959)
13. C. Kittel, W.A. Yager, and F.R. Merrit, Physica 15 256 (1949)
14. A.F. Kip, and R.D. Arnold, Phys. Rev. 75, 1556 (1949)
15. R.C. Le Craw, E.G. Spencer, and C.S. Pater, Phys. Rev. Rev. 110 1311 (1958)
16. C. Kittel, Phys. Rev. 110, 1295 (1958)
17. M.H. Seavey, and P.E. Tannenwald, Phys. Rev. Letters 1, 168 (1958)
18. T.G. Phillips, and H.M. Rosenberg, Reports Progress in Physics 29, 285 (1966)
19. K. Yosida, Progr. Theor. Phys. 7 , 425 (1952)
20. R.K. Wangsness, Phys. Rev. 97 831 (1955)
21. T.R. McGuire, Phys. Rev. 97 831 (1955)
22. A.J.E. Welch, P.F. Nicks, A. Fairweather and F.F. Roberts, Phys. Rev. 77 403 (1950)
23. S. Geschwind, and L.R. Walker, J. Appl. Phys. 30, 163 (1959)
24. J.F. Dillon, S. Geschwind, and V. Jaccarino, Phys. Rev. 100, 750 (1955)
25. G.P. Rodrigue, H. Meyer and R.V. Jones, J. Appl. Phys. 31, 3765 (1960)
26. L.R. Bickford, Phys. Rev. 76, 137 (1949)
27. W.A. Yager, J.K. Galt, F.R. Merritt and E.A. Wood, Phys. Rev. 80, 744 (1950)

GENERAL BIBLIOGRAPHY

Books

R.S. Alger,	"Electron Paramagnetic Resonance: Techniques and Applications", Interscience,London, New York,1968
B.I. Bleaney and B. Bleaney	"Electricity and Magnetism" 3rd edition. Oxford 1976 Oxford University Press, London
D.J.E. Ingram	"Spectroscopy at Radio and Microwave Frequencies" 2nd edition, Butterworths, London 1967
C.P. Poole	"Electron Spin Resonance. A Comprehensive Treatment on Experimental Techniques." Interscience, London, New York. 1967

Review Papers on Aspects of Ferromagnetic Resonance

Covalent Bonding and Magnetic Properties of Transition Metal Ions.
J.Owen, and J.H.M. Thornley, 1966, Reports on Progress in Physics
29 675 - 728.

Spin-Waves in Ferromagnets
T.G. Phillips, and H.M. Rosenberg, 1966, Reports on Progress in Physics, 29 285 - 332.

Anti-Ferromagnetic Resonance
T. Nagamiya, K. Yosida, and R. Kubo, 1955, Advances in Physics 4 1- 112

Ferromagnetic Resonance
G.T. Rado, 1961, J. Appl. Phys. 32 129 S. - 139 S.

APPENDIX A[1]

S.I. UNITS

Magnetic dipole moment — Am^2

Intensity of magnetization M — Am^{-1}

B field or magnetic flux density B — $NA^{-1}m^{-1}$ or T

H field or magnetic field strength H — Am^{-1}

Permeability of freespace $\mu_o = 4\pi \times 10^{-7}$ NA^{-2} or Hm^{-1}

Volume susceptibility $\chi = M/H$

Relative permeability $\mu_r = \mu/\mu_o = 1 + \chi$

e.m.u.

1 gauss = 10^{-4}T

1 oersted = $(10^3/4\pi)$ Am^{-1}

1 emu of magnetization M = 10^3 Am^{-1}

The magnetic permeability in e.m.u. is equal to the relative permeability μ_r defined above.
The value of the magnetic susceptibility (per unit volume) in e.m.u. is smaller than the SI value by a factor $1/4\pi$ (note that unit volume in e.m.u. is cm^3 and in S.I. is m^3)

[1] B. Bleaney and B.I. Bleaney, Electricity and Magnetism
3rd Edition, Oxford University Press 1976

APPENDIX B[2]

Basic Constants:

Speed of light	$c = 2.997925 \times 10^{8} ms^{-1}$
Fine structure constant	$\alpha = 7.29735 \times 10^{-3}$
Elementary charge	$e = 1.60219 \times 10^{-19} C$
Planck's constant	$h = 6.6262 \times 10^{-34} Js = 4.13517 \times 10^{-15} eVs$
	$\hbar = 1.0546 \times 10^{-34} Js = 6.5822 \times 10^{-16} eVs$
Avogadro's number	$N_o = 6.02205 \times 10^{23} mol^{-1}$
Electron rest mass	$m_e = 9.10954 \times 10^{-31} kg = 5.4859 \times 10^{-4} amu$
Magnetic flux quantum	$\Phi_o = 2.067851 \times 10^{-15} Tm^2$
Bohr magneton	$\mu_B = 9.27408 \times 10^{-24} JT^{-1} = 5.7884 \times 10^{-5} eVT^{-1}$
Nuclear magneton	$\mu_N = 5.0508 \times 10^{-27} JT^{-1} = 3.1525 \times 10^{-8} eVT^{-1}$
Magnetic moment of free electron	$\mu_e = 9.28483 \times 10^{-24} JT^{-1} = 1.00115964 \mu_B$

[2] see E.R. Cohen, Atomic Data and Nuclear Data Tables, 18, 588 (1976)

APPENDIX C Energy Equivalents

	J	eV	Hz	cm^{-1}	K	cal	T	kg	amu
1) 1 J =	1	$6,24146.10^{18}$	$1,50916.10^{33}$	$5,03403.10^{22}$	$7,24312.10^{22}$	$2,39006.10^{-1}$	$1,07827.10^{23}$	$1,11265.10^{-17}$	$6,7006.10^{9}$
2) 1 eV =	$1,60219.10^{-19}$	1	$2,41797.10^{14}$	$8,06547.10^{3}$	$1,16049.10^{4}$	$3,82933.10^{-20}$	$1,72759.10^{4}$	$1,78268.10^{-36}$	$1,07356.10^{-9}$
3) 1 Hz =	$6,62619.10^{-34}$	$4,13570.10^{-15}$	1	$3,33564.10^{-11}$	$4,79943.10^{-11}$	$1,58370.10^{-34}$	$7,14482.10^{-11}$	$7,37262.10^{-51}$	$4,43994.10^{-24}$
4) 1 cm^{-1} =	$1,98648.10^{-23}$	$1,23985.10^{-4}$	$2,99792.10^{10}$	1	1,43883	$4,74781.10^{-24}$	2,14197	$2,21026.10^{-40}$	$1,33106.10^{-13}$
5) 1 K =	$1,38062.10^{-23}$	$8,61708.10^{-5}$	$2,08358.10^{10}$	$6,95007.10^{-1}$	1	$3,29976.10^{-24}$	1,48868	$1,53615.10^{-40}$	$9,25098.10^{-14}$
6) 1 cal =	4,18400	$2,61143.10^{19}$	$6,31434.10^{33}$	$2,10624.10^{23}$	$3,03052.10^{23}$	1	$4,51147.10^{23}$	$4,65533.10^{-17}$	$2,8035.10^{10}$
7) 1 T =	$9,27410.10^{-24}$	$5,78839.10^{-5}$	$1,39961.10^{10}$	$4,66861.10^{-1}$	$6,71734.10^{-1}$	$2,21657.10^{-24}$	1	$1,03188.10^{-40}$	$6,21420.10^{-14}$
8) 1 kg =	$8,987554.10^{16}$	$5,60954.10^{35}$	$1,3565.10^{50}$	$4,52436.10^{39}$	$6,50979.10^{39}$	$2,14808.10^{16}$	$9,6910.10^{39}$	1	$6,02220.10^{26}$
9) 1 amu =	$1,49241.10^{-10}$	$9,31481.10^{8}$	$2,2523.10^{23}$	$7,51284.10^{12}$	$1,08097.10^{13}$	$3,5669.10^{-11}$	$1,60922.10^{13}$	$1,66052.10^{-27}$	1

Author index

Subject index